理工系の技術英語

論文の作成・発表に必要なスキル

Sonia Sharmin・奥 浩昭 著

Technical English for Science and Engineering

Writing and Presentation Skills

共立出版

Preface

The textbook **is intended for** undergraduate-3rd-year students or others to acquire skills for writing and presenting an academic paper, through the reading of texts written in easy English.

The book consists of two parts. Part 1 discusses basic skills needed to construct a paragraph, report, abstract, etc. Part 2 presents linguistic skills needed for improving writing and presentation of an effective academic paper. English texts on science and engineering are offered so that students can become familiar with a variety of disciplines and learn **relevant** vocabulary and words expressing logical connectives. Each chapter has exercises. In part 1 the exercises concern the structure of a paragraph, report, etc. In part 2, the exercises are focused on rhetorical functions and vocabulary. In both parts, students are given assignments concerned with writing paragraphs and reports.

For help with study, the Japanese translation of the texts, a grammar supplement, answers to the exercise questions, and discussion on word origins are given. A model paper is presented at the beginning of the book. It is hoped that as the students progress through each chapter, they will consult the model paper to better understand the principles being discussed. Another important aim of this textbook is to expand the student's basic science-related vocabulary. On the right-hand margin, the relevant Japanese words and meanings are therefore provided to facilitate the students' easy understanding. Section numbers for the relevant grammar explanation and word origins are also given there.

The textbook was designed and written by a scientist who is an active researcher and whose first language is English, with the help of a Japanese educator who, specializing in linguistics, taught English to engineering students. The authors hope that, through the study of the text, students will be able to write and present a paper with confidence.

June, 2024　Sonia Sharmin

be intended for … ～を目指している
the reading of ⇒ the –ing of ～の形をとる（例：The Making of the English Bible）
on ⇒ G7.6
relevant 関連する

specializing in linguistics ⇒ G16.3
through the study of the text ⇒ G16.3

まえがき

本書は，平易な英語で書かれた文章の読解を通して，学術論文を書き，発表するのに必要なことがらを身につけることを目指した，学部3年生他を対象としたテキストです。

本書は2部で構成されています。第1部はパラグラフやレポート，アブストラクトなどを書くのに必要な基本的なスキルを扱います。第2部では，より効果的な論文を書き，発表するのに必要な言語表現技術を学びます。英文テキストは理工系の諸分野を紹介しており，学習者は理工系の幅広い分野の教養を身につけながら，関連する語彙や論理表現を学ぶことができます。各章に練習問題がついています。第1部ではパラグラフやレポートの構造が把握できているかを問い，第2部では論理表現や語彙などの理解を問います。第1部でも第2部でも，パラグラフやレポートを書く課題が与えられます。

学習の参考に，テキストの日本語訳と文法，練習問題の解答，語源の紹介を加えました。また，モデルペーパー（模範論文）を本書の最初に載せました。各章を学びながらこのモデルペーパーを参照し，学習内容の理解を深めてほしいと思います。また，テキストを読みながら科学に関する語彙力をつけてほしいと思います。そのため，みなさんが容易に理解できるよう，ページの右側に関連する単語と意味を載せています。文法と語源で扱われる関連する語句のセクション番号も右側に載せました。

本書は，英語を母語とする理工系の先端分野の研究者が，理工系の大学で英語教育に携わった日本人教員の協力を得て，企画，執筆しました。本書の学習を通し，研究成果を自信をもって英語でまとめ，発表する力を身につけてほしいと願っています。

2024年6月　Sonia Sharmin

読者のみなさんへ（Prefaceの補足）

　本書は，Sharmin 先生が執筆した英文テキスト Chapter1-30 に，奥が右側の側注（語彙，文法・語源の重要語についての説明）と章末の日本語訳を追加し，さらに，巻末付録 Ｇ 文法，付録 Ｗ 語源，参考文献（の一部）を新たに執筆しました。文法と語源の習得によって，英語の構造の理解と英単語への関心が深まることを期待しています。

　この本の学習目標は，英語での論文作成・発表のスキルを身につけることです。この目標を達成するために，以下にみなさんへのエールを記します。

　翻訳ソフトの質が画期的に向上し，日本語の文章がたちどころに，ほぼきちんとした英語へ機械翻訳されるようになった現在，皆さんの中には，英語力の向上にさほど真剣に取り組む必要はないのでは，と思う方もいるかもしれません。その気持ちはよく理解できます。ただ，将来科学者や技術者として学会や各種打ち合わせなどで外国の人と接する時，翻訳ソフトに頼るというわけにはいきません。相手の話す英語を理解し，それに対し時間をおかずに自分の考えを述べなければなりません。また，研究を進める際には，世界共通語である英語で書かれた論文を短時間で読み，内容を把握しなければなりません。したがって，英語を読み，書き，話す（その前提として聞いて理解する）という基本的な学習が不可欠です。

　テキストを読む際に，みなさんに意識してほしいことが２つあります。

・ひとかたまりのテキスト（たとえば一つのパラグラフ）を，辞書を引かずに読み，大まかな内容の理解に努める。
・少なくとも２回は英文に向かう。わからない単語や表現が出てきても，すぐに右側の側注を見たり辞書を引いたりしない。わからない単語や表現をすべて調べるのではなく，パラグラフの内容理解を妨げている単語や表現についてのみ，意味を文脈からある程度推測したうえで，右側の側注をみる。

　なお，日本語訳については直訳を心がけていますが，場合によっては意味が通ることを優先し，いわゆる意訳となっている箇所があります。具体例を挙げます。翻訳ソフトの訳は，ややわかりづらいのではないでしょうか。

原文：A cause in one sentence or clause can be linked to an effect in the following sentence or clause by a connector.（24.1 (3)）
DeepL：ある文または節の原因は，コネクターによって次の文または節の結果と結びつけることができる。
Google：ある文または文節の原因は，接続子によって次の文または文節の結果にリンクされます。
日本語訳：文と文，あるいは節と節の間の因果関係はコネクタ（接続語）によって表されます。

　細かな点ですが，英文と日本語訳の書式にも，以下の違いがあることを断っておきます。

・英文では，パラグラフの最初を一字空け（indent）していません。現在の英文作成においては，それが主流です。一方日本語訳では，通例通りパラグラフの最初は一字空けています。
・英文では，見出し語句の一文字目を大文字にしています。一方，日本語訳では固有名詞以外は，すべて小文字にしています。たとえば，表の記述に違いがみられます。英文で So となっているとろを，日本語訳では so としています。

2024 年 6 月
奥浩昭

Model paper

A model paper has been supplied in this part so that you can learn how to write a scientific article. Please read the relevant chapters in Part 1 to understand more about the specific sections.

Title: Effect of cleaning with nanobubbles **on** resistivity of silicon wafer

effect of A on B　A が B に及ぼす影響

Abstract:

Any nanoscale contaminants on silicon wafer surfaces, such as quartz, iron, stainless steel, or aluminum, can lead to errors in the extremely precise etching procedure required to print computer circuits on them. A unique method of surface cleaning uses nanobubbles, which are microscopic gaseous (usually air) cavities found at the liquid/solid interface. To determine the effectiveness of nanobubble cleaning, we measured the resistivity of silicon wafers both before and after cleaning. A well-known method for producing nanobubbles, the so-called alcohol-water exchange operation, which involves first coating the surface with alcohol (ethanol or propanol) and then flushing it with water, was used in the study. After cleaning, it was discovered that the silicon wafer had a resistivity of 90 ohm-cm, compared to 55 ohm-cm for the wafer covered in particles. Our results demonstrate that the spherical polystyrene nanoparticles produced on the surfaces are effectively removed by the alcohol-water exchange process, which causes the creation of nanobubbles.

Introduction:

Any nanoscale impurities on silicon wafer surfaces, such as quartz, iron, stainless steel, or aluminum, which are all ubiquitous in a manufacturing environment, may create flaws in the incredibly precise etching process needed to print computer circuits on them. Many cleaning cycles are required to achieve the desired level of cleanliness, which frequently involves the use of environmentally toxic chemicals. However, nanobubbles, which are nanoscopic gaseous (typically air) cavities located at the liquid/solid interface, are a novel technique to clean surfaces. Contaminants can be gently removed and washed away by allowing them to develop on a wafer surface [1].

Surface tension draws a bubble inward, and the smaller the bubble, the higher the internal air pressure must be to keep it from collapsing. The pressure inside a bubble with a diameter of 100 nm should theoretically be at least five times that of the surrounding liquid, easily high enough to dissolve the gas. Therefore, such small bubbles should not exist. Surprisingly, Japanese researchers observed such minuscule bubbles on a silicon surface covered with water in 2000 [2]. The so-called alcohol-water exchange procedure, in which the surface is first covered with alcohol (ethanol or propanol), and then flushed with water, is now a well-known technique for creating nanobubbles. The discovery has been confirmed for both hydrophilic bare silicon surfaces and a hydrophobic silicon substrate [3]. The higher gas solubility in alcohol in comparison to water may be responsible for the formation of nanobubbles in the alcohol-water exchange process. When water is used in place of alcohol, gas is removed from the liquid, resulting in oversaturation and the formation of nanobubbles. Furthermore, the process is an exothermic reaction that generates heat, which likely contributes to the development of nanobubbles.

According to previous research, a growing nanobubble spreads horizontally on the surface (in other words, a nanoscopic gas layer accumulates on the surface), on a time-scale of seconds, increasing its surface coverage and then developing in height. Any contaminating nanoparticles on the surface are thought to become detached by this action [4]. In our study, we have measured resistivity of the silicon wafers before and after being cleaned with nanobubbles to understand their cleaning efficacy.

surprisingly ⇒ G 5.3

alcohol–water exchange procedure ⇒ G 16.1

hydrophilic ⇒ W 18

be responsible for ⇒ G 16.9.3

resulting ⇒ G 11.2

Method: In our experiments, a bare silicon wafer (with a native dioxide layer on top) was cleaned with acetone and then with ethanol. Following each treatment, blow drying with nitrogen gas was done. In a spin-coating machine, a drop (0.5 mL) of nanoparticle solvent (polystyrene latex spheres, diameter 100 nm) was injected onto the substrate. The particles were then spin-distributed across the surface (speed 6100 rpm, duration 10 s).The sample was naturally dried at room temperature during the spinning process.

a spin–coating machine ⇒ⓖ16.1

spin–distributed ⇒ⓖ16.1

A fluidic cell was connected with a syringe-pumped inlet and outlet for the ethanol-water exchange process to form nanobubbles at a solid-water interface while simultaneously scanning with an atomic force microscope (AFM). After placing the sample beneath the fluidic cell, the alcohol/water exchange process was carried out on the AFM stage [1]. As a result, AFM measurements in liquid could be performed immediately following the fluidic process that generates the nanobubbles.

a syringe–pumped inlet and outlet ⇒ⓖ16.1

following ⇒ⓖ11.2

The resistivity of the silicon wafer was measured using the four-probe method.

using ⇒ⓖ11.2
four–probe method ⇒ⓖ16.1

Results: AFM was used to first image the particle-covered dry surface. On the surface, randomly scattered nanoparticles could be seen. The AFM imaging (in liquid) was performed at the same surface location as the dry measurement. After the fluidic process, nanobubbles were discovered, while all nanoparticles had been removed from the surface (see Figure 1).

to first image ⇒ⓖ12.8

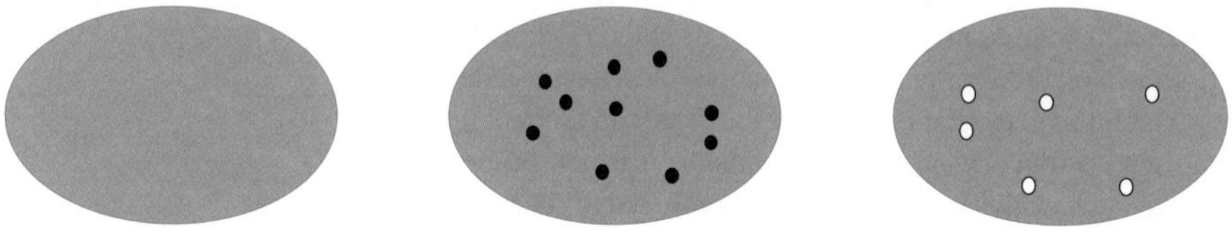

Silicon wafer

Silicon wafer with nanoparticles

Silicon wafer with nanobubbles

Figure 1: After cleaning with nanobubbles (white dots), no nanoparticles (black dots) were evident on the surface of the silicon wafer

After cleaning, the resistivity of the silicon wafer was found to be 90 Ω cm, while the resistivity of the particle-covered wafer was 55 Ω cm. The bare silicon wafer used at the beginning of the experiment had a resistivity of 100 Ω cm.

Discussion: The evidence of nanobubble cleansing is supported by the resistivity measurement. Furthermore, it demonstrates that nanobubbles are **stable**, which aids in the prevention of pollutants resettling.

stable 安定している

Other nanobubble generation mechanisms (e.g., electrolysis, temperature increase, gas supersaturation), or a combination with conventional cleaning methods may point to additional research on improving cleaning efficiency. In real-world industrial applications, other types of particles, such as aluminum, aluminum oxide, gold, or copper, are more commonly encountered. Our future work will be focused on the best ways of removing such materials.

e.g. ⇒ⓖ16.9.6

real–world industrial applications ⇒ⓖ16.1

Conclusion: The resistivity of the wafer before and after cleaning was compared in experimental studies of surface cleaning with nanobubbles on a bare silicon wafer. Our findings show that the alcohol-water exchange process, which results in the formation of nanobubbles, efficiently removes the spherical polystyrene nanoparticles deposited on the surfaces.

References:

1. Yang, S., & Duisterwinkel, A. (2011). Removal of nanoparticles from plain and patterned surfaces using nanobubbles. Langmuir, 27(18), 11430-11435.
2. N. Ishida, T. Inoue, M. Miyahara, and K. Higashitani, "Nano Bubbles on a Hydrophobic Surface in Water Observed by Tapping-Mode Atomic Force Microscopy," Langmuir 16, 6377 (2000).
3. Tyrrell, J. W., & Attard, P. (2001). Images of nanobubbles on hydrophobic surfaces and their interactions. Physical review letters, 87(17), 176104.
4. Fang, C. K., Ko, H. C., Yang, C. W., Lu, Y. H., & Hwang, S. (2016). Nucleation processes of nanobubbles at a solid/water interface. Scientific reports, 6(1), 1-10.

モデルペーパー

　理工系の学術論文の書き方を身に着けてもらえるよう，モデルペーパーを示します。関連する第1部の章を読み，理工系英語の特定の内容の理解を深めてください。

タイトル：ナノバブルによる洗浄がシリコンウェーハの抵抗率に及ぼす影響

概要

　石英や鉄，ステンレス鋼，アルミニウムなどのナノスケールの汚染物質がシリコン・ウェーハの表面にあると，コンピューター回路の印刷に必要となる非常に精密なエッチング手順にエラーが発生する可能性がある。表面洗浄というユニークな方法ではナノバブルを使用する。ナノバブルは，液体と固体の界面に見られる微細な気体（通常は空気）の空洞である。ナノバブル洗浄の効果を調べるため，洗浄前と洗浄後のシリコンウェーハの抵抗率を測定した。本研究では，ナノバブルを発生させるためのよく知られた方法，いわゆるアルコール－水交換操作（まず表面をアルコール（エタノールまたはプロパノール）でコーティングし，次に水で流す）が用いられた。洗浄後，粒子で覆われたウェーハの抵抗率は55 Ω・cmであるのに対し，シリコンウェーハの抵抗率は90 Ω・cmであったことが判明した。この結果は，表面に生成した球状のポリスチレンナノ粒子が，ナノバブルを生成するアルコール－水交換プロセスによって効果的に除去されることを示している。

序論

　石英や鉄，ステンレス鋼，アルミニウムといった，製造環境に広く存在するシリコンウェーハ表面上のナノスケールの不純物により，コンピュータ回路の印刷に不可欠な，精密なエッチングプロセスに不具合を生じるおそれがある。洗浄サイクルを繰り返すことで清浄度のレベルを十分に高めなければならないが，洗浄の際，環境に有毒な化学物質を使用することがよくある。しかし，表面を洗浄するのにナノバブルが考えられる。液体と固体の界面に配置されたナノスコピックな気体（通常は空気）の空洞であるナノバブルは，表面をきれいにする新しい技術である。ウェーハ表面に付着させることで汚染物質を徐々に取り除き，洗い流す，というものだ [1]。

　表面張力によりバブルが内側に引き寄せられるため，バブルが小さいほど，崩壊しないように内部空気圧を高める必要がある。直径100 nmのバブル内の圧力は，理論的には周囲の液体の少なくとも5倍と，ガスを溶解するのに十分なものでなければならない。そのため，そのような小さなバブルは存在しないはずだが，驚くべきことに，2000年に日本の研究者が，水で覆われたシリコン表面にそのような小さなバブルを観察した [2]。最初に表面をアルコール（エタノールまたはプロパノール）で覆い，次に水で洗い流す，いわゆるアルコール－水交換手順は，ナノバブル作成のよく知られた手法である。この発見は，親水性の裸のシリコン表面と疎水性のシリコン基板のいずれでも確認されている [3]。水と比較してアルコールのガス溶解度が高いため，アルコール－水交換プロセスにおいてナノバブルが形成される可能性がある。アルコールの代わりに水を使用すると，液体からガスが除去され，過飽和状態になり，ナノバブルが形成される。さらに，このプロセスは熱を発生させる発熱反応であり，ナノバブル発生の要因となる可能性がある。

　従来の研究によれば，ナノバブルは数秒で表面に水平に広がり（つまり，ナノスコピックガス層が表面に蓄積し），その表面被覆率を高めたうえで高さを増す。そしてこの作用により，表面の汚染ナノ粒子は分離すると考えられている [4]。本研究では，ナノバブルで洗浄する前後のシリコンウェーハの抵抗率を測定し，それらの洗浄効果を調べた。

方法

　本実験では，裸のシリコンウェーハ（上部に天然の二酸化炭素層がある）をまずアセトンで，次にエタノールで洗浄した。二つの洗浄の後，窒素ガスによるブロー乾燥を行った。スピンコーティング機で，ナノ粒子溶媒（直径100 nmのポリスチレンラテックス球）を1滴（0.5 mL）基板に注入した。次に，粒子を表面全体にスピン分散させた（速度6100 rpm，持続時間10秒）。スピンプロセス中，サンプルを室温で自然乾燥させた。

　流体セルは，エタノール－水交換プロセス用のシリンジポンプ式の入口と出口に接続され，原子間力顕微鏡（AFM）で同時にスキャンしながら，固体－水界面でナノバブルを形成した。流体セルの下にサンプルを置いた後，AFMステージでアルコール/水交換プロセスを実行した [1]。その結果，液体中のAFM測定を，ナノバブルを生成する流体プロセスの直後に実行できた。

　シリコンウェーハの抵抗率は，4プローブ法を使用して測定した。

結果

　AFMを使用して，最初に，粒子で覆われた乾燥した表面を画像化した。表面には，ランダムに散乱したナノ粒子が見られた。AFMイメージング（液体中の）は，乾量測定と同じ表面位置で行った。流体プロセスの後すべてのナノ粒子を表面から除去する間に，ナノバブルが発見された（図1を参照）。

シリコンウェーハ

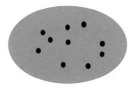
ナノ粒子の付着した
シリコンウェーハ

ナノバブルの付着した
シリコンウェーハ

図1：ナノバブル（白点）での洗浄の後，シリコンウェーハの表面にナノ粒子（黒点）は見られなかった。

　洗浄後，シリコンウェーハの抵抗率は 90 Ωcm，粒子で覆われたウェーハの抵抗率は 55 Ωcm であった。実験開始時に使用した裸のシリコンウェーハの抵抗率は 100 Ωcm だった。

考察
　ナノバブルによる洗浄の効果が，抵抗率測定によって明らかとなった。さらに，ナノバブルが安定して存在することがわかり，汚染物質の再発生を防止しうることがわかった。
　他のナノバブル生成メカニズム（たとえば，電気分解，温度上昇，ガス過飽和），または従来の洗浄方法との組み合わせにより，洗浄効率の改善に関するさらなる研究が期待される。実際の製造環境では，アルミニウムや酸化アルミニウム，金，銅などの他の種類の粒子がより多く使用されている。今後の研究により，そのような材料を取り除く最良の方法を提案したい。

結論
　単純なシリコンウェーハ上のナノバブルを用いた表面洗浄の実験により，洗浄前後のウェーハの抵抗率を比較した。その結果，ナノバブルを生成するアルコール－水交換プロセスが，表面に堆積した球状ポリスチレンナノ粒子を効率的に除去することがわかった。

参考文献
1. Yang, S., & Duisterwinkel, A. (2011). Removal of nanoparticles from plain and patterned surfaces using nanobubbles. Langmuir, 27(18), 11430-11435.
2. N. Ishida, T. Inoue, M. Miyahara, and K. Higashitani, "Nano Bubbles on a Hydrophobic Surface in Water Observed by Tapping-Mode Atomic Force Microscopy," Langmuir 16, 6377 (2000).
3. Tyrrell, J. W., & Attard, P. (2001). Images of nanobubbles on hydrophobic surfaces and their interactions. Physical review letters, 87(17), 176104.
4. Fang, C. K., Ko, H. C., Yang, C. W., Lu, Y. H., & Hwang, S. (2016). Nucleation processes of nanobubbles at a solid/water interface. Scientific reports, 6(1), 1-10.

Table of contents

目次

Part 1

Basic skills for research writing and presentation

Preface-Part 1

Students can achieve success when they use written and oral communication effectively. In Part 1 of this book, students will be able to learn the basic skills needed for effective written and oral science communication. Each lesson has relevant vocabulary and **linking words** highlighted, as well as exercises at the end of each chapter to check the students' understanding. For science students, a great deal of work has to be done around report and research article construction and so this is discussed in detail. Tasks teaching how to **paraphrase** so as to avoid **plagiarism** are also presented.

There are numerous advantages to effective written science communication, such as, increasing clarity, avoiding misunderstandings, **enhancing** productivity and reducing errors and mistakes. For communicating clearly and easily, first the goal has to be identified and stated clearly so that the reader understands what needs to be done or researched. Also, for the reader to acquire interest in the research presented, the tone used throughout the writing must be appropriately enthusiastic.

Students must also be able to express their academic needs and goals verbally. Oral communication includes live presentations, video presentations, and interviews, as well as visual aids such as handouts and PowerPoint presentations. Knowledge of or use of vocabulary, grammar, and sentence structure, as well as strong comprehension skills, are all components of oral language proficiency. Many people are terrified of speaking in public, especially in large groups. However, good preparation can **alleviate** these fears while also laying the groundwork for an effective presentation. The chapters in Part 1 thus cover research, preparation and presentation of various tasks required at university and future science-related occupations.

Since research builds on previous research, all academic work requires accurate referencing. As a student, the general principles to follow when citing sources must be understood and **precautions** taken to avoid plagiarism. Written resources are nowadays easily accessible on the internet, and students may not be aware of what constitutes plagiarism. Proper citation of sources prevents plagiarism and establishes the author's and paper's credibility.

linking word 連結語

paraphrase 言い換え
plagiarism 剽窃，盗作

enhance 高める

alleviate 和らげる

precaution 予防策（複数形で用いられることが一般的）

第 1 部
学術論文の作成・発表に必要な基本的スキル

第 1 部への序

　しっかりとした論文を書き口頭での発表がうまくできるようになれば，将来研究者や技術者として成功を収めることができます。本書の第 1 部では，効果的な論文の書き方や口頭発表の基礎を身につけることができます。各章では，重要な語彙や連結語が強調されています。また，章末の練習問題では，学んだことが理解できているか確認することができます。理工系の学生は，論文やレポートの作成に多大の労力を要しますので，本書はその作成方法を詳細に述べています。剽窃を避けるための言い換えの書き方も示しています。

　効果的な論文を書くことには，明晰さを増す，誤解を生じさせない，生産性を高める，誤りを減らすなど多くのメリットがあります。明確でわかりやすいコミュニケーションを成り立たせるためには，まず研究の目標を定め，述べることです。そうすれば読者は，その研究の目的が遂げられるのに必要なことを理解できます。また，読者がその研究に興味と関心をもてるよう，論文の最初から最後まで，著者は冷静かつ熱意をもって書き進めなければなりません。

　学生はまた，研究のニーズと目標を口頭で的確に述べられるようにならなければなりません。口頭発表には対面やビデオでの発表，インタビューがあり，また，配布資料やパワーポイントな

どの視覚資料の使用を伴うことがあります。強い理解力に加え，語彙や文法，文の構造の理解と使用が口頭発表の熟達には必要となります。人前で話すことを多くの人が不安に思います。ましてや多くの人の前ではなおさらです。しかし，しっかり準備をして臨めば不安は和らぎますし，効果的なプレゼンテーションの基礎固めができます。したがって第 1 部では，大学または将来研究職や技術職に就いた際に求められる，さまざまなタスクの研究，準備，発表を扱っています。

　学術研究は先行研究の成果のうえに成り立っていますので，論文や口頭発表などの学術研究は正確な参考文献を載せなければなりません。先行文献を引用する際に守らなければならない一般的な原則を理解し，剽窃を避ける細心の注意を払う必要があります。インターネットで資料が簡単に手に入る今日，剽窃が何かを知らずにいる学生もいるかもしれません。文献を適切に引用すれば剽窃に陥ることはなくなり，著者と論文への信頼が得られます。

a report and paragraph
⇒🅖2.5.2

Chapter 1

Format of a report and paragraph

Students need to learn the basics of presenting their research in a well-organized manner, employing sections and headings to ensure that the information is readily accessible and comprehensible. This is called a report. Original reports, when their findings signify a noteworthy progress in comprehending a crucial issue and carry broad and immediate significance, are termed articles or research papers.

The whole report is divided into several sections, and each section is divided into paragraphs. To understand how to write a report, we need to first understand how to write a paragraph showing its significance to the report as a whole and then how to link it to the next section/paragraph.

each ⇒🅖3.8
we ⇒🅖3.1
to first understand ⇒🅖12.8

1.1 Paragraph

Each paragraph expresses a complete thought.

1.1.1 Structure of a paragraph

The paragraph should contain a topic sentence, **preferably** at the beginning, which discusses the main **hypothesis** presented in the paragraph. It should answer the questions *who, what, when* and *where*. The rest of the paragraph should **be devoted to** supporting the hypothesis, i.e., giving reasons for your opinion. This means that the remainder explains how or why or both. Finally, a concluding statement should be written **relating** the significance of what has been written.

the ⇒🅖2.4
preferably できれば ⇒🅖
5.4
hypothesis 仮説
be devoted to … ～に捧げら
れる，～に向けられる
i.e. ⇒🅖16.9.7
relate 述べる

(1) Topic sentence

The topic sentence can be introduced at the start of a paragraph or placed at the end to provide a summary of the entire concept. The topic sentence **clarifies** the entire paragraph for the reader.

The remaining sentences should provide additional information, elucidation, or proof for the central idea.

Characteristics of a topic sentence:

clarify 明確にする（clear
にする）

- · The topic sentence needs to **cover** all the information presented in the paragraph.
- · The topic sentence must be **specific** to the paragraph.
- · The topic sentence must announce exactly what the paragraph is about – no more, no less.

Make sure to revisit your paragraph, ideally after a while. Verify that each sentence aligns with the main topic sentence.

cover 取り上げる，述べる
specific 特定の，一般的で
ない

(2) Conclusion

A brief passage does not necessarily need a closing remark, but in the case of a lengthier one, it is advisable to include a conclusion rather than an abrupt ending. The significance of a conclusion lies in its role as the last impression conveyed to the reader. Hence, the conclusion should encapsulate the key message you wish the reader to retain. A concluding statement can take one of the following approaches:

- · Reinforce the primary point for emphasis, achieved by rephrasing or rewording the topic sentence.
- · Recapitulate the information to provide a concise review or clarification.

In a concise paragraph, the summary may closely align with the main point.
· Relate the significance or importance of what was written. Future effects and actions can be stated.

1.1.2   Paragraph writing skills and vocabulary

When writing a paragraph, you must choose a logical order in which to present information so that one idea leads to another. The paragraph may be organized according to any one of the following concepts:

in which to present ⇒G12.6
any ⇒G3.8

(1) Time (chronology)
Moving from the past to the current situation to possible future outcomes. Each part **in turn** is the cause of the **subsequent** part.

chronology 時間，時系列
from A to B to C ⇒G7.3
current ⇒W4
in turn そして，次に，今度は ⇒G7.7
subsequent それに続く

(2) Space
Moving from one location to another location and so on.

(3) Logic
Moving from the general to the specific or moving from the specific to the general or moving from the simple to the complex.

(4) Comparison and contrast
A discussion of the similarities and differences between two or more items is frequently required in scientific writing. There are two basic ways of comparison:

: すなわち ⇒G16.2.1

Method 1: Describing the **attributes** of the items **one by one**.
Method 2: Comparing all items **one** characteristic **at a time**.

attribute [ˈætrɪbjuːt] 特性
one by one 次々と
one ... at a time 1 回につき 1 個を
transition word 転換語
specify 明確に述べる

Signpost or **transition words specify** how two ideas are related. Common transition words and phrases such as *also, furthermore*, *in the same way, likewise, moreover, similarly,* etc. express comparison. Transition words and phrases that show contrast include *although, but, despite, even though, however, in contrast, in spite of, nevertheless, on the contrary, on the other hand, yet,* etc.

(5) Cause and effect
Is caused by, is induced by, is produced by, happens because, is due to etc. introduce the reason for something. Effects are introduced with the words *produced, as a result, hence, caused, induced, consequently, thus, therefore, for this reason, so, because of this* etc.

(6) Examples
Introducing examples is a good way to illustrate the different aspects of the idea of the paragraph more clearly. Transition words and phrases that express examples are *for instance, for example, to illustrate, such as, like* etc.

1.2   Format of a report

A scientific report or paper usually follows a certain order with certain sections. The organization of the paper is quite **vital** for ease of use for both the reader and the writer. The sections of a paper are:

vital 重要である

(1) **Abstract:** Summarizes the whole report
Verbal forms: Simple present and simple past tenses

(2) **Introduction:** Explains briefly the purpose in writing the report
Verbal forms: Simple **present tense**

present tense 現在時制，現在形

(3) **Background:** Presents what is already known and unknown to explain why the investigation was necessary
Verbal forms: Simple **past tense**

past tense 過去時制

(4) **Method:** Shows the process carried out in the investigation

Verbal forms: Simple past tense

(5) Result: Presents the evidence or data discovered
Verbal forms: Simple past tense, also passive voice to have the focus on the action rather than the **agent**

agent 行為者

(6) Discussion: Discusses the results presented, providing statements, **speculations** and criticisms
Verbal forms: For speculation, *must have, cannot have, may* or *might*. For criticism, use *should* or *ought to.*

speculation 予想

(7) Conclusions: Gives an **overall** opinion based on all the evidence
Verbal forms: Should + perfect **infinitive** to indicate criticism, and must + perfect infinitive to indicate certainty

overall 全体的な

infinitive 不定詞

(8) Recommendations: Suggests the actions others should take as a result of the report
Verbal forms: modal *should*

(9) Supplement: provides additional tables, diagrams, references and documents

---

Exercises

Q Read the following text, and answer the questions.

Sometimes, planets in our solar system are classified according to their position relative to the Earth. Because their orbits are closer to the sun than Earth's, Mercury and Venus **are referred to as** inferior planets. Those (planets) whose orbits are larger are called superior planets. Recent missions to Mars have not turned up definite proof of life. Planets may also be classified according to their size or mass. Because their surfaces are compact and rocky, comparable to Earth's geography, the planets Mercury, Venus, Earth, and Mars are referred to as **terrestrial**. The outer planets Jupiter and Saturn are gas giants, while Uranus and Neptune are ice giants. Jupiter is the largest followed by Saturn, and both are **substantially** larger than Uranus and Neptune.

their ⇒ⓖ 3.9

are referred to as … ～と呼ばれる

terrestrial 陸の

substantially かなり

i. Which of the following is the best topic sentence for this paragraph?
(a) Planets may be classified according to their sizes and masses.
(b) Planets may be classified according to their sizes and orbital distances.
(c) Many planet classification schemes have been proposed over the years.
(d) The planets in our solar system are classified according to their position relative to Earth.

ii. Which sentence does not belong in the paragraph?

iii. Write a paragraph in approximately 150 words on a world-changing breakthrough in science.

第1章　レポートとパラグラフのフォーマット

　学生は，自分の研究を整理して発表するための基本を学ぶ必要があります。その際，セクションや見出しを用いることで，受け手が簡単に情報にアクセスでき，理解しやすいようにします。これをレポートと呼びます。レポートは，その発見が重要な問題を理解するうえでの顕著な進歩を意味し，広範で直接的な意義をもつとき，論文または研究論文と呼ばれます。

　レポートはいくつかのセクションからなり，各セクションは複数のパラグラフ（段落）から構成されます。レポートの書き方を学ぶにはまず，レポートの中で重要な意味をもつパラグラフの書き方とパラグラフ相互の関連づけを理解する必要があります。

1.1　パラグラフ

各パラグラフはまとまった意見を述べます。

1.1.1　パラグラフの構造

パラグラフでは，そのパラグラフで提示される仮説を述べるトピック・センテンスをできれば第一文に置かなければなりません。これには「いつ，どこで，だれが，なにを」に対する答えを示す情報が含まれています。パラグラフのそれ以降の文章はその仮説の根拠すなわち理由の提示に充てられます。すなわち，残りの部分で「どのように」と「なぜ」のどちらか，あるいは両方を説明します。そしてパラグラフの最後に結論が述べられ，これまで論じてきたことの重要性が示されます。

(1)　トピック・センテンス

トピック・センテンスは，パラグラフの最初に置くことも，最後に置いてコンセプト全体の要約を示すこともできます。それにより，読者はパラグラフ全体の内容を明確に理解できます。

残りの文章は，要点に対する追加の情報または説明，証明を提供する必要があります。

トピック・センテンスには，以下の特徴があります。

・パラグラフで提示するすべての情報を提示する
・パラグラフに関わることだけを述べる
・パラグラフの主題のみに言及する

理想的には，しばらくしてからパラグラフを見直すようにしましょう。各文章が，メインのトピック・センテンスと一致していることを確認してください。

(2)　結論

短い文章であれば必ずしも結びの言葉は必要ありませんが，長い文章では，唐突に終わるのではなく結びの言葉を入れるのが望ましいです。結論の重要性は，読者に伝える最後の印象としての役割にあります。そのため，結論の部分では読者に心に留めておいてほしい重要なメッセージを凝縮して示す必要があります。結論の文章は，以下のいずれかのアプローチをとることができます。

・トピック・センテンスを言い換えたり，表現を変えたりして，要点を強調する
・簡潔なレビューや説明を提供するために，情報を要約する

簡潔なパラグラフでは，要約は要点と密接に一致することがあります。

1.1.2　パラグラフライティングに必要なスキルと語彙

パラグラフを書く際には，一つのアイデアが次のアイデアにつながるよう，情報を論理的に順序立てて提示する必要があります。パラグラフは次の要素の中のいずれかに沿って構成することができます。

(1)　時間

過去から現在，起こりうる未来の結果へ展開します。それぞれの部分は順番に後続の部分の原因となります。

(2)　空間

ある場所から別の場所などへ展開します。

(3)　論理

一般から具体へ，あるいは具体から一般へ，また単純から複雑へ，と展開します。

(4)　比較と対照

科学論文ではしばしば，2つ以上の項目の類似性と相違に着目した議論が必要となります。比較には大きく2つのやり方があります。

方法1：対象の特性を一つひとつ記述する
方法2：すべての項目を一度に一つの特性ごとに比較する

道標となるあるいは転換を表す語は，2つのアイデアがどのように関連しているかを示します。also, furthermore, in the same way, likewise, moreover, similarly などは転換語です。対照を表現する語には although, but, despite, even though, however, in contrast, in spite of, nevertheless, on the contrary, on the other hand, yet などがあります。

(5)　原因と結果

is caused by, is induced by（誘発される）, is produced by, happens because, is due to などは理由を示す表現です。結果は次のような語（句）で示されます：as a result, hence[*1], caused, induced,

*1 セクション 6.4.1 参照.

consequently, thus, therefore, for this reason, so, because of this

⑹　**例示**
パラグラフで示された意見のもつさまざまな側面を読者が容易に理解できるよう，例を挙げるのが効果的です。その際，for instance, for example, to illustrate, such as, like などの転換語句が用いられます。

1.2　レポートのフォーマット

　科学レポートや科学論文は通常一定の順序に従い，一定のセクションをもっています。論文のこの構成は読者にとっても著者にとっても使いやすく，きわめて重要です。論文は以下の要素で構成されます。

⑴　**アブストラクト**：論文全体の要旨
　　動詞の形態：現在形と過去形

⑵　**序論**：論文の目的を簡潔に記す
　　動詞の形態：現在形

⑶　**背景**：解明されたこととされていないことを示し，本研究の必要性を説明する
　　動詞の形態：過去形

⑷　**方法**：行った研究の実施過程を示す
　　動詞の形態：過去形

⑸　**結果**：得られた結果としての証拠やデータの提示
　　動詞の形態：過去形。なお，実験そのものに焦点を当てるため，受動態を用いる

⑹　**考察**：得られた結果を考察し，その結果を，肯定的断定・可能性の示唆・批判的評価で述べる
　　動詞の形態：可能性の示唆は must have, cannot have, may or might を用い，批判的評価は should または ought to で表す

⑺　**結論**：証拠に基づき総合的見解を述べる
　　動詞の形態：批判的評価には should + have + 過去分詞を，確実性の示唆には must + have + 過去分詞を用いる

⑻　**推奨**：本論文の結果を受けての読者への示唆
　　動詞の形態：助動詞の should を用いる

⑼　**補足**：その他の表や図，参考文献，資料を載せる

Chapter 2
Steps to better reading

Understanding the steps of scientific research can assist in focusing a scientist's inquiry, and working through his/her observations and data to obtain the best possible result.

Stages of Scientific Research

(1) Choosing a topic
(2) Identifying a problem
(3) Generating ideas and methods
(4) Selecting an approach
(5) Deciding how to solve the problem
(6) Choosing the best solution of those available
(7) Developing a plan and time line
(8) Carrying out the plan on schedule
(9) Evaluating good and bad points
(10) Arriving at conclusions
(11) Sharing the results with other people
(12) Clearly expressing all ideas
(13) Correctly and clearly presenting materials and information

After choosing a topic, the first step in a research project is to obtain a good understanding of the issues involved, and for that, reading of relevant papers, journals, books, and magazines are the only way. However, such a huge amount of **literature** has to be tackled in a correct manner; otherwise, it will **overwhelm** the hopeful scientist at the beginning.

a good understanding ⇒ G 2.5.1
literature 文献
; ⇒ G 16.2.1
otherwise ⇒ W 5
overwhelm 圧倒する，意欲を削ぐ

2.1 Preview

Studies indicate that skilled and self-assured readers utilize multiple approaches to comprehend the meaning of a written passage. Good readers begin by previewing the text, i.e., looking over the entire passage for a few moments.

a few ⇒ G 3.7

(1) Consider the title: When previewing, the first thing to do is read the title. Titles **prompt** the reader to consider what the author wishes to **address**, as well as announcing the topic.

prompt 促す
address 取り組む

(2) Prediction: Prediction is a fundamental reading skill that is used to guess how a passage will **unfold**. Readers can use information from a text, relate it to their prior knowledge and use it to make logical predictions before and during reading.

prediction 予測 ⇒ W 6
unfold 展開する

> Synonym of predict: *foretell, forecast, foresee*

synonym 同義語

> Difference between *anticipation* and *prediction*: "Anticipation" is used when we expect the event to happen soon or at a specific time. "Prediction" is employed when we expect the event to happen further in the future or at an unknown time.

(3) Read the first paragraph: Some writers may state at the outset what they hope to **convey** or why they are writing. Some writers attempt to capture the reader's attention by posing a difficult question. In any case, the opening paragraph aids in understanding the purpose of the text. In research papers, the Abstract at the beginning is a paragraph that

convey 伝える

summarizes the whole paper.

(4) Read the final paragraph: Some writers end by repeating the main idea, while others draw a conclusion or summarize.

(5) Glance through: First, skim through to get a quick sense of the information. Then scan to find the relevant information.

Skimming strategies: Skimming and scanning are reading skills that most students use frequently in their native language. Skimming is the process of quickly reading a text to grasp the main idea. Scanning is the process of quickly viewing a text in order to find key terms, phrases, or information.

Read only the first and last sentences of each paragraph when skimming an article. During skimming, you are reading to understand the **gist**, i.e., the main idea of the text. You do not need to read every word. **Instead**, search for:

· Concerns
· Using words in a specific order, such as *first* and *second*.
· Reasonings explained with **clue** words like *however, but, now, to explain.*
· Conclusions and suggestions.

gist [dʒɪst] 要旨
instead そうではなく

clue 手がかりとなる，結びつける

2.2 Read for comprehension

Read the whole text and try to understand the meaning. Single words lack significant meaning on their own. Readers need to see meaningful combinations of words and phrases <u>to **eventually** comprehend</u> the passage completely.

eventually 最後には
⇒[G]5.4
<u>to eventually comprehend</u>
⇒[G]12.8
<u>is being discussed</u> ⇒[G]10.2

Theme and rheme: The theme of a sentence is the topic or what <u>is being discussed</u>, and the comment (rheme or focus) is what is said about the topic. The theme comes at the beginning of a sentence and includes all the words **up to** the first verb.

up to ... ～まで

2.3 Understanding the paragraphs

To understand the paragraph and what an author is **implying**, the reader needs to be able to find out some clues. The following **tips** should help:

imply ほのめかす
tip 情報，ヒント

(1) Identify the topic sentence: The topic sentence, which is frequently the first sentence of a paragraph, contains the primary concept of the paragraph. It can, however, appear at the end of a paragraph. The other sentences support, develop and explain the main idea. In some cases, there may be no topic sentence in the paragraph. These paragraphs, which usually create a mood or feeling rather than presenting information, are not commonly found in scientific texts.

(2) Understand the purpose of the paragraph: <u>Each</u> paragraph has a different purpose. The purposes may be:

<u>each</u> ⇒[G]3.8

(A) inform - to give knowledge or information;
(B) define - to describe the meaning or character of something;
(C) describe - to give details about something;
(D) persuade- to convince someone to do or think about something;
(E) explain - to make something clear or to give the reason for something;
(F) entertain - to interest or please;
(G) illustrate - to make clear or to tell a story <u>using</u> drawings, pictures, examples or comparisons;
(H) compare or contrast - to discover the similarities and differences between two or more people or things.

<u>using</u> ⇒[G]11.2

2.4　Organize the facts

The reader needs to organize the facts, since comprehending how all the pieces of information come together to convey the message is the true challenge. A good reader can discover the author's plan or motive by looking for clues or signal words in the text that may **reveal** the author's motivation. Therefore, the last step is to identify the relevant details and then look for connections, relationships and **inferences** so that the reader can get to the **ultimate meaning**.

reveal 明らか

inference 推論
ultimate meaning 究極の意味，著者の考え

Exercises

Q1 In the following situations, which is more useful, scanning or skimming?

(a) Reading the **Table of Contents** of a book to decide whether it **covers** the area you are seeking.

table of contents 目次
cover 取り上げる

(b) Consulting an Index to locate the pages where Newton's First Law is referenced.

(c) Reviewing a chapter of a book to determine if it includes any mention of Albert Einstein.

(d) Examining an article to assess whether it potentially contains relevant information for your research project.

Q2 Read the following paragraph carefully and answer the question:

There have been three great **innovation**s in the development of communication. First was the invention of writing which changed the world. Written information can be referred to easily as it is recorded. Second, printing permitted the mass production of books in the 15th century. It was possible to publish and share scientific results and experimental data with a wide audience because of this novel ability. Finally, no other invention has had as much impact on every part of our lives as electronic technology, which allows people to communicate over long distances (TV, Internet, phone). Whereas before people had to rely on letters, they can now maintain a relationship with a vast group of people by modern digital communication technology.

innovation ⇒Chap. 19

What is the main idea of the text?

(a) The significance of effective communication
(b) The three groundbreaking developments in communication
(c) The methods employed by individuals for communication
(d) The creation of the Internet

第 2 章　より良い読み方へのステップ

　科学研究の手順を理解していれば，研究テーマから逸れることなく，実験の観察やデータに基いた最適の結果が得られます。

科学研究の手順

(1) トピックの選択
(2) 問題の把握
(3) 解決に至るアイディアと方法の創出
(4) 問題へのアプローチの確定
(5) 解決法の決定
(6) 最適解の選択
(7) 計画の確定と解決までにかかる時間の設定
(8) スケジュールに沿った計画の実行

⑼ 成果となお残る問題の評価
⑽ 結論の提示
⑾ 研究結果の開示
⑿ 研究で明らかになったすべてのアイデアの明記
⒀ 研究で用いた資料・材料・情報の正確で明確な公開

取り上げるトピックが決まれば，研究プロジェクトにおいてまず課題をよく理解し，関連する論文やジャーナル，本，雑誌などを読まねばなりません。しかし，膨大な文献に対する対応を誤ってはなりません。さもないと将来ある若い研究者は入り口でくじけてしまいかねません。

2.1　全体像の把握

研究によれば，熟練した自信のある読者は，書かれた文章の意味を理解するために複数のアプローチを利用していることがわかっています。そのような読み手はテキストをプレビューする，つまり，テキスト全体を時間をかけずに読み，内容を読み取れるのです。

⑴ **タイトルに注目する**：全体像の把握には，まずは，タイトルを読むことからです。タイトルはトピックを告知するだけでなく，著者が何を訴えたいのかを読者に考えさせます。

⑵ **予測しながら読む**：予測は，基本的な読解力でテキストの展開を予想します。書かれていることと自分の事前の知識を重ね合わせて，テキストの次の部分で何がいわれるのかを，次の部分に入る前に，また，その部分を読みながら予測するのはとても大切です。

> predict（予測する）の同義語：foretell, forecast, foresee, make a prognosis

> anticipation と prediction の違い：anticipation は，ある事象が近いうちに特定の時間に起こりそうなときに用いられるのに対し，prediction は，ある事象の発生が遠い先か，いつと特定できないときに用いられます。

⑶ **第一パラグラフを読む**：最初のパラグラフで著者がある問題についての見解を示したり，取り上げる理由を述べることがあります。最初に難しい問いを投げかけることで読者を引きつける，ということもありえます。いずれにせよ，最初のパラグラフはテキストの目的の理解を助けてくれます。論文では，アブストラクトが論文全体の要約のパラグラフになります。

⑷ **最後のパラグラフを読む**：最後のパラグラフに主要な考えや結論，要約を述べることがあります。

⑸ **まず全体をざっと読む**：まず全体をざっと読み，論文への第一印象をつかみます。そのうえで，重要な情報を探します。

> スキミング：スキミング，スキャニングは流し読みのことで，母語のリーディングではふつうに行っています。スキミングは速読による主要な考えの把握です。一方スキャニングは，速読による重要表現や情報の抽出です。
> 　スキミングの際には各パラグラフの最初と最後の文だけを読み，要旨，すなわち主要な考えの把握に努めてください。一字一句を読むのではなく，次の表現に注目しましょう。
>
> ・トピック
> ・first, second などの順番を表す語
> ・however, but や now（さて），explain などの語を用いた論の展開
> ・結論や提案

2.2　全体を理解して読む

テキスト全体を読み，その意味の理解に努めてください。単語は単一で置かれると，それ自体で重要な意味をもちません。意味のある語や語句のつながりに注目し，テキスト全体の趣旨を理解するようにしてください。

> テーマとレーマ：言語学では，文のテーマはトピックあるいは話題の対象であり，レーマ（コメント，フォーカス）はトピックについての意見です。テーマは文頭に置かれ，最初の動詞までのすべての単語を含みます。

2.3　パラグラフの理解

　パラグラフの内容と著者の意図を理解するには，次のような「読む手がかり」を見つける必要があります。

(1)　**トピック・センテンスを探す**：トピック・センテンスはパラグラフの基本概念を含んでおり，そのパラグラフの第一文に来ることが一般的ですが，最後の文に来ることもあります。その他の文は，主要な考えを支持し，発展させ，説明します。トピック・センテンスのないパラグラフもあります。その種のパラグラフは，情報を提示せずに，筆者の感情や思いを伝えるはたらきをしており，自然科学のテキストに出てくることはあまりありません。

(2)　**パラグラフの目的を理解する**：それぞれのパラグラフには異なる役割があります。次のような目的が考えられます。

　(A)　知らせる：知識や情報を与える。
　(B)　定義する：扱う対象の意味や性質を述べる。
　(C)　描写する：扱う対象について詳細に述べる。
　(D)　説得する：著者の主張・見解を読者に納得させる。
　(E)　説明する：扱う対象を明確にする，あるいは主張の理由を明確に述べる。
　(F)　楽しませる：読者の関心をひきつける。
　(G)　図示する：図や絵，例，比較などにより理解を助ける。
　(H)　比較，対比する：2 つ以上の物の共通する点と違いを見つける。

2.4　事実の整理

　すべての情報がどのように組み合わさってメッセージを伝えているのかを理解することが本当の課題のため，読者は事実を整理する必要があります。テキスト内の手がかりとなる語，著者の動機を明らかにしている語を探すことで，読者は著者の計画や動機を知ることができます。したがって最後のステップは，テキストの理解に重要な細部の表現を確認し，表現間のつながりや関係，推論を探して，テキストの意味にたどり着くことです。

Chapter 3

Writing an Introduction – Expressing purpose

The Introduction section of a research paper or report gives an overall view and provides a brief, non-technical summary. By reading the Introduction section **alone**, the reader should be capable of discovering whether the research or the theory put forward in that article can be beneficial to him/her.

… alone 〜だけで

3.1 Stages of an Introduction

For a scientific paper or report, usually two or three paragraphs to introduce the research topic are needed. The introduction should contain the following four elements:

(1) Some context to **orient** the readers who are less familiar with the research topic should be offered and <u>the significance of the research work established</u>. Another not unusual approach used to interact with the reader is to set up the need for further research on the selected **subject** matter.

(2) The introduction should summarize the **relevant literature** and explain <u>any</u> important technical words which will be used extensively throughout the paper.

(3) The writer's own opinion or a possible answer to his/her research question should be expressed clearly in the Introduction.

(4) Finally, the reader needs to be informed about what constitutes the different sections of the research paper.

orient 導く
the significance of the research work established ⇒Ⓖ 16.8
subject トピック
relevant literature 関連の ある文献
any ⇒Ⓖ 3.8

3.2 Expressing purpose

A purpose is usually expressed by an **adverbial clause** which <u>typically</u> begins with the **conjunctive phrases** *so that* and *in order to*. Modal **auxiliary verbs** such as *should, might, may, will, can* and *could*, usually follow.

Examples:
- Low density materials need to be used so that they may be treated as **equivalent**.
- The new material has to be tested carefully in order to be used as <u>a room-temperature superconductor</u>.

adverbial clause 副詞節
typically 通常 ⇒Ⓖ 5.4
conjunctive phrase 接続詞句
auxiliary verb 助動詞
equivalent 同等
a room-temperature superconductor ⇒Ⓖ 16.1

Difference between *may* and *might*: These days, you may use *may* and *might* <u>interchangeably</u>; however, there is a slight distinction between the two. *May* is used to express what is possible, real, or can be real. *Might* expresses a stronger sense of doubt or a <u>contrary-to-reality</u> **hypothesis**. They both indicate that something is **feasible**; however, "something that **may** happen" is much more likely than "something that might take place".

interchangeably どちらも 同義で使える ⇒Ⓖ 5.4
contrary-to-reality ⇒Ⓖ 16.1
hypothesis 仮説
feasible 実行可能な

3.2.1 Subordinate clauses expressing purpose

Subordinate clause: A clause that does not form a simple sentence by itself but adds extra information to the sentence.

Subordinate clauses of purpose answer the query *why* or *what … for*. The use is presented in the **main clause**. There are three possible ways for using a subordinate clause to **convey intent**:

subordinate clause 従属節

main clause 主節
convey intent 意図を伝える

(1) (in order/so as) to + infinitive

Example:
- In order to locate the position of the event, only three **space coordinate**s need to be known.

space coordinate 空間座標

(2) (in order/so) that + verb

Example:
- The inner surface of the shell must bear a **positive charge** so that the total charge enclosed within the shell is balanced and amounts to zero.

positive charge 正電荷

(3) for + noun/pronoun + to-infinitive

Example:
- For you to go forward we need you to pass this obstacle.

Negative forms of clauses expressing a purpose

Examples:
i. The loops are widely separated so as not to interfere with one another.
ii. In order not to fall behind, we have adopted a process of continuous improvement.

3.2.2   Difference between subordinate clauses expressing purpose

(1) We use *to, in order to, with a view to* and *so as to* + infinitive when the subjects or topics of both clauses are identical.

Examples:
- I study hard to excel in my exams.
- They met in order to share information about the project.
- He studied hard so as to excel in the final exams.

(2) We use *in order that* or *so that* when the subjects of the two clauses are different.

Examples:
- The **dipole** can be rotated about an axis in order that the angle is changed.
- The wheel is inverted so that it is seen to be rotating **clockwise**.

dipole 双極子
clockwise 時計回りに⇒Ⓦ5

Here, the subjects of the main and subordinate clauses are *wheel* and *we,* respectively.

Lest: to avoid the risk of – A word expressing purpose

Example:
- Unique resources were used **lest** there be copyright **infringement**.

lest ... ～するといけないので
infringement 侵害

Exercises

Q1 Correct the following sentences by changing the word forms in the brackets:

i. The metal brushes slide against the plane (for make) contact with each other.

ii. The **rim** contains lead (for to make) it stronger.

rim へり

Q2 Rewrite the following sentences <u>using</u> the words in brackets to show purpose.

using ⇒Ⓒ 11.2

i. She studied hard since she wished to understand the complex concepts. (in order to)

ii. The **refrigerator** was **modified**. The energy loss is minimum. (so that)

iii. The north pole of the magnet faces the approaching north pole. They will **repel** one another. (so as to)

refrigerator 冷蔵庫
modify 改造する（文法用語としては「修飾する」）
repel 退ける ⇒Ⓦ7

第3章　序論を書く―目的を述べる―

　学術論文やレポートの序論の部分では全体像を示し，平易な要約が簡潔に記されます。序論の部分を読むだけで，読者はその論文やレポートが自分に役立つものなのか判断できるはずです。

3.1　序論の組み立て

　学術論文やレポートでは通常，研究のトピックを示す2個ないし3個のパラグラフが必要です。序論には以下の4つの要素を含まなければなりません。

(1)　研究テーマをあまり知らない読者をうまく導くような文脈を提供し，研究の意義を確かなものにする。読者との対話でよく用いられるもう一つのアプローチは，選んだテーマに関するさらなる研究の必要性を説くことである

(2)　序文では関連文献を要約し，論文全体で広く用いる重要な専門用語を説明する必要がある

(3)　序論では，著者自身の意見や，研究課題への可能な解答を明確に表現する

(4)　最後に，研究論文の他のさまざまなセクションについて読者に知らせる必要がある

3.2　目的を述べる際に用いられる表現

　目的は so that や in order to などの接続詞句と，その後に来る should, might, may, will, can, could などの助動詞を伴う副詞節で表されます。

例：
- Low density materials need to be used **so that** they **may** be treated as equivalent. (低密度の素材は，同等のものとして扱えるように用いなければならない。)
- The new material has to be tested carefully **in order to** be used as a room-temperature superconductor. (室内の温度で用いる超伝導体として使用できるよう，この新しい材料は入念なチェックを受けなければならない。)

---

may と might の違い：今では may と might はよく同義で使われていますが，両者にはわずかながら違いがあります。may は可能，事実，事実でありうる場合に用いられるのに対し，might は事実や可能に対する話し手（書き手）の疑念が強いことを表します。あるいは，現実に反する仮定を表します。どちらも実現可能な事態を表しますが，"something that **may** happen" は "something that **might** take place" よりその可能性がはるかに高いのです。

3.2.1　目的を表現する従属節

従属節：単独の文としてではなく，文に新たな情報を加える節のことです。

　目的を表す従属節は why や what … for などの問いに答えます。答えは主節で示されます。目的の表現には3つのタイプがあります。

(1)　**(in order/so as) to 不定詞**

　　例：
　　- **In order to locate** the position of the event, only three space coordinates need to be known. (事象の位置を知るには3つの空間座標（3次元）の値がわかるだけで十分だ。)

(2)　**(in order/so) that ＋ 動詞**

　　例：
　　- The inner surface of the shell must bear a positive charge **so that** the total charge enclosed within the shell is balanced and amounts to zero. (殻の内面は，殻内に囲まれた総電荷のバランスが取れてゼロになるように，正の電荷を帯びていなければならない。)

(3)　**for ＋ noun/pronoun ＋ to 不定詞**

　　例：
　　- **For you to go** forward we need you to pass this obstacle. (前進するにはこの障害を突破しなければならない。)

目的を表す従属節の否定形

　例：
ⅰ．The loops are widely separated **so as not to** interfere with one another. (互いに影響を与えないようループとループには十分な間隔が置かれている。)

ⅱ．**In order not to** fall behind, we have adopted a process of continuous improvement. (遅れを取らぬ

よう，不断に上達する道を選んだ。）

3.2.2　目的を表す従属節間の違い

(1)　to, in order to, with a view to, so as to + 不定詞を用いるのは，両方の節の主語やトピックが同じ場合です。

例：
- I study hard **to excel** in my exams.（私は試験で優秀になるために一生懸命勉強する。）
- They met **in order to** share information about the project.（プロジェクトに関する情報を共有するため集まった。）
- He studied hard **so as to** excel in the final exams.（彼は最終試験で優秀な成績を収めるために一生懸命勉強した。）

(2)　2 つの節の主語が異なる場合，in order that や so that を用います。

例：
- The dipole can be rotated about an axis **in order that** the angle is changed.（軸を中心に双極子を回転させることで，双極子の角度を変えることができる。）
- The wheel is inverted **so that** it is seen to be rotating clockwise.（輪が反転するのは時計回りに回転していることを示すためである。）

この文の主節と従属節の主語はそれぞれ，wheel と we です。

Lest：〜の危険を回避するため。目的を表す語。

例：
- Unique resources were used **lest** there be copyright infringement.（著作権の侵害を避けるため，独自のリソースを使用した。）

Chapter 4

Describing method

In a research paper there are different stages. After the Introduction, where the background and context of the research is discussed, the Method section is placed. This section describes how the research was done. The Method section is sometimes termed **Procedure**.

procedure 手続き

 Stages in a research report:

(1) Introduction
(2) Method
(3) Results
(4) Discussion
(5) Conclusion

4.1 Process of writing the Method section

Any scientific investigation or experiment is first planned and then repeated and **modified**. The conditions of the experiment have to be carefully controlled to acquire reliable results. The Method section has thus to be very carefully and clearly written.

modify 修正する，修飾する

(1) List all the steps in the research process.
(2) Establish the steps in **chronological order**.
(3) Confirm that every step is distinct and separate, ensuring nothing has been overlooked.

chronological order 時系列

The sentences should be rather short, and only one step of <u>the procedure should be addressed</u> in one sentence. In this way, the section will become clear to the readers. All necessary details, such as quantities, measurements, and times should be written so that the same experiment may be undertaken by another researcher and reproduced.

<u>the procedure should be addressed</u> ⇒ **G** 16.5

Nouns used in the Method section:
Hypothesis, procedure, observation, experimentation, investigation, analysis, test, equipment, device, **apparatus**, design, set-up, accuracy, temperature, material, technique

hypothesis 仮説
apparatus [ǽpəˈrætəs] 装置

4.2 Expressions used in the Method section

4.2.1 Past tense

When writing the Method section, past events are being related, so usually the simple past tense is used. The following ways of expression are commonly found in this section:

(1) by + -ing form:

Example:
・The pole was secured **by attaching** it to the ground with five chains.

(2) (by) using / by means of + noun:

Examples:

· The pole was secured by using five chains to attach it to the ground.
· The temperature of the sample was increased by means of the **thermal energy** produced.

thermal energy 熱エネルギー

4.2.2 Active and passive voice

In a research report it is expected that the usage of **pronouns** should be minimized as much as possible. On the other hand, the **active voice** is preferred as it makes the meaning of the sentences easier to understand. This **contradiction** between the usage of personal pronouns and the usage of active voice has to be resolved by the **judicious** usage of the **passive voice** in different sentences.
The active and passive voices have been discussed in detail in the Supplementary Grammar section 10.

pronoun 代名詞
active voice 能動態

contradiction 矛盾 ⇒W6
judicious 賢明な
passive voice 受動態

4.2.3 Using *when* and *as*

There is a slight difference between the two words *when* and *as*, which are often found in the Method part of a paper. They both express actions performed **in sequence**. However, *when* is used when the two actions happen immediately one after the other.

in sequence 連続して

Example:
· When minuscule oil droplets are introduced, some of them acquire an electric charge.

When the two actions happen **simultaneously**, **i.e.** at the same time, *as* is used.

simultaneously 同時に
i.e. すなわち ⇒G16.9.7

Example:
· As the piston moves, air is **compressed**.

compress 圧縮する（com-
= completely press, 圧力
を加える）
succession 連続

However, if it is sometimes difficult to distinguish whether the actions happen separately but in a very fast **succession** or simultaneously, either *as* or *when* can be used.

Example:
· When/As the field is removed, the **induced dipole moment** disappears.

induced 誘導された
⇒W10
dipole moment 双極子モーメント

4.2.4 Words and phrases signaling sequence

The following words and phrases are commonly used to highlight the **transitions** between the different stages of the research procedure:

transitions 移行

at the same time, immediately afterwards, simultaneously, then, after this, soon afterwards

Examples:
· At the same time, heat energy is generated.
· Both wavelengths were then used simultaneously.
· After this, the particle followed an **elliptical** path.

elliptical 楕円形の

Exercises

Q1 Fill in the gaps with the correct form of the words in **parentheses**. These sentences may be used in the Method section of a research paper.

parenthesis（単数形）括弧

i. Drills were used to _____ (make) holes. Then the holes were _____ (fill) with **resin**.

resin [ˈrezn] 樹脂

ii. The powder _____ (flow) from the pipe into the **reservoir**.

reservoir [ˈrezərvwɑːr]
貯水池
alternator [ˈɔːltərneɪtər]
交流機，発電機

iii. The **alternator** was _____ (drive) by the engine.

iv. The air was then _____ (cool).

v. The motion was _____ (control) by a switch.

Q2 Write the **verbs** which are related to the following nouns used to describe actions:　　**verb** 動詞形
　　Observation, experimentation, prediction, investigation, interpretation, construction

Q3 Write six sentences using each of the nouns in (Q4).

Q4 Think about an experiment you can set up <u>where</u> wind power is converted into elec-　　<u>where</u> ⇒先行詞は an exper-
trical power. Draw a simple **diagram** and then describe the setup. You are free to include　　iment
any materials and objects that you might wish.　　**diagram** ダイアグラム

第4章　方法を述べる

　研究論文の執筆には，いくつかの段階があります。研究の背景や目的を記した序論の後に，方法のセクションを書きます。ここでは研究の手順が示されます。方法セクションを手順と呼ぶこともあります。

研究レポートの構成：
　(1)　序論
　(2)　方法
　(3)　結果
　(4)　考察
　(5)　結論

4.1　方法の書き方

　科学的な研究や実験ではまず構想が立てられ，それに基いた実験が繰り返されたり，修正が加えられたりします。実験を行う環境や条件を正確に整えることで信頼できる結果が得られます。したがって方法セクションは，以下に注意して論理的にかつ明確に記されなければなりません。

　(1)　研究プロセスのすべてのステップを列挙する
　(2)　時系列に手順を確立する
　(3)　すべてのステップが明確で独立していることを確認し，見落とされていないことを確認する

　また，各文は短く，また1つの段階のみを記さなければなりません。そうすることで，研究の方法が読者に明確に示されます。量，計測，時間などの必要な詳細をすべて明確に示すことで，他の研究者も同じ実験を正確に繰り返すことができます。

方法セクションで用いる名詞：hypothesis, procedure, observation（観察），experimentation（実験），investigation（調査），analysis（分析），test（テスト），equipment（装置），device（装置），apparatus, design（デザイン），set-up（セットアップ，段取り），accuracy（正確さ），temperature（温度），material（材料），technique（技術）

4.2　方法セクションで用いられる英語表現

4.2.1　過去形

　方法セクションでは行った事象に言及するので，ふつう過去形が用いられます。よく用いられるのは次のような表現です。

(1)　by + ing（動名詞）：

例：

・The pole was secured **by attaching** it to the ground with five chains.（地面に5本の鎖を取りつけることで，柱を固定した。）

(2)　(by) using / by means of + noun：

例：

・The pole was secured **by using** five chains to attach it to the ground.（5本の鎖を用い，地面に取りつ

けることで，柱を固定した。)
・The temperature of the sample was increased **by means of** the thermal energy produced.（生じた熱エネルギーを用いて，サンプルの温度を上げた。）

4.2.2　能動態と受動態

研究レポートでは代名詞の使用は極力回避されます。一方で能動態の使用が好まれます。その方が文意が理解しやすいからです。代名詞の回避と能動態の使用との間の矛盾は，文によって受動態をうまく使いわけることで解決しなければなりません。

能動態と受動態については付録🄶10 で扱います。

4.2.3　when と as の使用

方法セクションでよく用いられる when と as にはわずかながら違いがあります。どちらも連続した出来事を表しますが，2 つの出来事に前後関係があるときには when を用います。

例：
・**When** minuscule oil droplets are introduced, some of them acquire an electric charge.（微小な油滴が導入されると，その一部が電荷を帯びる。）

2 つの出来事が同時に起こる場合には as を使います。

例：
・**As** the piston moves, air is compressed.（ピストンの動きに合わせて空気は圧縮される。）

しかし，複数の行為が別々に，しかしほぼ連続してあるいは同時に起きているのかはっきりしない時には，どちらも使えます。

例：
・**When/As** the field is removed, the induced dipole moment disappears.（磁場がなくなると，誘起された双極子モーメントは消える。）

4.2.4　連続を表現する語や語句

以下の語句を用いて，研究手順の異なる段階間の移行を強調します。

at the same time（同時に）, immediately afterwards（その後すぐに）, simultaneously（同時に）, then（そして）, after this（その後）, soon afterwards（その後しばらくして）

例：
・**At the same time**, heat energy is generated.（同時に熱エネルギーも発生する。）
・Both wavelengths were then used **simultaneously**.（そして，両方の波長が同時に用いられた。）
・**After this**, the particle followed an elliptical path.（この後，粒子は楕円軌道をたどった。）

Chapter 5

Presenting results

Research is primarily based on an extensive variety of activities. The consequences of research activities need to be scientifically measured and then stated. The Results section of a paper or report expresses what has been discovered as a result of study and experimentation. Tables, charts, and graphs organize information so that we can understand and compare easily.

5.1 Process of writing the Results section

The most important part of writing the Results section is deciding first what consequences are **relevant** to your findings, even though the Results section only presents the findings themselves. Afterwards, the results need to be organized according to the answers to the goals, hypotheses or questions which were set at the start of the paper. This is usually done in science **disciplines** with the help of figures and tables. **Additionally**, all essential negative results have to be mentioned.

relevant 関連のある

discipline 学問分野
additionally さらに ⇒ G5.3

5.2 Language used to describe facts

From an English point of view the important thing in writing the results of research is reporting them sincerely and completely. If the **referees** of the paper cannot understand the outcomes of the study, then this contribution to modern research may become lost.

In the Results section, facts and data are presented with the help of **diagrams**, tables, and graphs. The present tense is used when referring to any table or figure.

Adjectives are usually used to add some extra information. There are two kinds of adjectives: opinion adjectives and fact adjectives. Opinion adjectives express what a person thinks of someone or something, whereas fact adjectives provide information about age, size, shape, origin, texture, and a variety of other factors. If there are both opinion and fact adjectives the opinion adjective is placed before the fact adjective.

referee 査読者

diagram ダイアグラム

adjective 形容詞

Example:
· We could obtain quite excellent, <u>noise-free data</u>.

noise-free data ⇒ G16.1

5.3 Words and phrases to describe data properties

When writing about data, i.e., the results obtained from a study or experiment, the scientist has to reveal what the relationship is between the different facts being presented. It is not possible to just present a string of numbers, instead what type of event the data is indicating must be clarified in the text. Additionally, it needs to be **unambiguously** explained how and why such data are obtained at a specific condition and what <u>is being affected</u> as a result.

The following words and phrases are used in academic writing to describe data, and any relevant trends in a certain time period.

decline, *drop*, *fall*, ***fluctuation***, ***gentle***, *gradual*, *increase*, ***level off***, *peak*, *rapid*, *reach a peak*, ***remain constant***, ***remain stable***, *rise*, *sharp*, *slight*, *slow*, ***steep***

unambiguously 曖昧でなく
is being affected ⇒ G10.2
decline 減少する
fluctuation 変動
gentle ゆるやかな
level off 横ばいになる
remain constant 一定である
remain stable 安定している
steep 急, 険しい

Chapter 5

5.4   Comparison of data

We usually try to show a positive or negative **correlation** between two sets of data to further   correlation 相関
clarify the relationship.

Comparative words and phrases
- *similar to, alike, equally, in the same way, similarly, likewise, in an identical manner, the same, quite the same, almost the same, very much the same*
- *exactly, precisely, just the same*
- *to have much **in common**, to resemble, to be different from, less than, **markedly**, some-   **in common** 多くの共通点
  what, slightly, far*   がある
                                                                                                     markedly 著しく

Examples:
- The first is similar to the one used in this part.
- They all look alike.
- Even to a penguin, all penguins look the same.
- All states are equally probable.
- Similarly, we can say that energy is stored in a magnetic field.

5.5   Useful phrases

The following phrases are commonly used to draw the reader's attention to the visuals.

- The figure refers to ...
- This graph shows ...
- On the graph, it is to be noted ...   is to ⇒Ⓖ12.7
- Here a comparison between ...
- The graph represents ...
- As you can see, the main ...

When showing results during a presentation, the following phrases are helpful in drawing the
attention of the audience:

- We must focus our attention on ...
- Let's look more closely at ...
- Let me draw your attention to ...
- I'd like to show you ...
- As you can see from this graph ...

5.6   Example of a Results section

This is an example of a Results section.

In an experiment, to determine **deformation** under force, 15 metals were assessed for ri-   deformation 変形 ⇒Ⓦ8
gidity using horizontal and vertical forces **respectively**. The Results section for such an   using ⇒Ⓖ11.2
experiment may be written as follows: **respectively** それぞれ ⇒Ⓖ
Mean **rigidity** values for each metal and each condition are reported in **Table** 1. **Figure** 1   16.9.4
shows how gold and silver are deformed with an increasing force. The breaking point can   **rigidity** 剛性
be determined to be at force of 40 and 55 N respectively. It can be seen that vertical forc-   **table** 表
es produce more deformation on the metals than horizontally applied forces.   **figure** 図

Exercises

Q1 Fill in the gaps with the correct form of the words in parentheses. These sentences may be used in the Results section of a research paper.

   i. No **celestial** object has ever been discovered spinning faster than the rate calculated by this _____ .(analyze)

   celestial 天体の

   ii. This engine is **currently** being _____ for use in cars and spacecraft. (develop)

   currently 現在

   iii. The cost of producing the extremely low temperatures needed to generate the effect **impeded** the technological _____ of superconductivity. (advance)

   impeded 妨げる

   iv. The law of energy conservation explains this _____ fact. (experiment)

   v. The graph shows the results of three trials _____ a particular spring-box system. (involve)

Q2 Rewrite the following sentences so that they mean the same using the comparative pattern with the word in parentheses.

   i. **Saturn** is not as warm as the Earth because it is farther away from the Sun.
   ≒ Earth is _____ warmer because it is nearer to the Sun than Saturn. (a little)

   Saturn 土星

   ii. The alpha particle is significantly lighter than the gold **nucleus**.
   ≒ The gold nucleus is _____ than the alpha particle. (heavy)

   nucleus 核

Writing

Q3 Consider that a **rigidity test** was performed by you on a material to understand its deformation under a force. Using at least one graph, write a paragraph presenting the results of such a test. You may consider any material, value, conditions etc. **as needed**.

   rigidity test 剛性試験

   as needed 必要に応じて

第 5 章　結果の提示

　研究はさまざまな活動を通して行われます。そしてその結果は，科学的に客観性をもって計測され，述べられなければなりません。論文やレポートの結果セクションは，研究や実験の結果明らかになったことを記します。表や図，グラフの使用により，結果を整理して読み手（聴き手）に伝えることができます。

5.1　結果を書く手順

　結果セクションを書くに当たって最も重要なのは，結果セクションでは結果そのものを提示するだけであっても，発見したことに関連する研究結果をまず決めることです。そのうえで，得られたデータ（結果）を，論文の最初に記した，研究の目的や仮説，問いに対する答えとして提示するのです。理工系の分野では図や表を用いて示します。本質的な否定的結果が得られたとしても，それもすべて記さなければなりません。

5.2　事実の記述に用いられることば

　結果の記述に当たって重要なのは，事実をありのまま，すべて提示することです。事実をありのまま，すべて提示することをしなかったがために論文の査読者が研究結果を理解できないとしたら，せっかくの研究も学問の前進に貢献できなくなってしまいます。

　結果セクションでは，事実とデータを表すのに，ダイアグラムや表，グラフを用います。表や図の説明には現在形が使われます。

　形容詞は付加情報を表します。筆者の意見を表す形容詞（opinion adjective）と事実を述べる形容詞（fact adjective）からなります。前者は得られた事実やデータなどへの評価を示し，後者は年齢，大きさ，形，産地，生地や質感などさまざまな要素を表します。両者が同時に用いられる場合には，筆者の意見を表す形容詞，事実を述べる形容詞の語順をとります。

例：
・We could obtain quite excellent, noise-free data. （かなり質のよい，ノイズのないデータが得られた。）

5.3　データ特性の表現に用いられる語句

事実（＝研究や実験の結果得られた結果）を記述する際，個々のデータ間の関係を明らかにする必要があります。単にデータを羅列するのではなく，個々のデータが示している事象を明示的に示さねばなりません。さらに，曖昧さのない表現で，どうやって，なぜ，どのような条件下でこのデータが得られるのか，その結果どのような影響が生じているのかを記述しなければなりません。

学術論文では，以下の語（句）を用いて，データや特定の時間に得られた関連する傾向を記します。

decline, drop, fall, fluctuation, gentle, gradual, increase, level off, peak, rapid, reach a peak, remain constant, remain stable, rise, sharp, slight, slow, steep

5.4　データの比較

2 つのデータ間の相関の有無を示し，両者の関係を鮮明に示します。

比較の際に用いる表現

- similar to, alike, equally, in the same way, similarly, likewise, in an identical manner, the same, quite the same, almost the same, very much the same
- exactly, precisely, just the same
- to have much in common, to resemble（似ている）, to be different from, less than, markedly, somewhat（いくらか）, slightly（少々）, far（はるかに）

例：
- The first is **similar to** the one used in this part.（1 つ目は，このパートで使用したものと似ている。）
- They all look **alike**.（みんなそっくり。）
- Even to a penguin, all penguins look **the same**.（ペンギンはペンギン。）
- All states are **equally** probable.
- **Similarly**, we can say that energy is stored in a magnetic field.（同様に，エネルギーは磁場に蓄えられているといえる。）

5.5　有用な表現

図やグラフなどのビジュアルなものに読者の関心を引くのに以下の表現がよく用いられます。

- The figure refers to ...（この図は〜を指している。）
- This graph shows ...
- On the graph, it is to be noted ...（グラフ上で注目すべき点は以下の通りです。）
- Here a comparison between ...（ここでは〜との比較を示します。）
- The graph represents ...（グラフは〜を表しています。）
- As you can see, the main ...（ご覧のとおり，主な〜）

結果を示すのに，聴き手の注意を引くのに以下の表現が有効です。

- We must focus our attention on ...
- Let's look more closely at ...（もっと詳しく見てみましょう。）
- Let me draw your attention to ...（〜に注目してみましょう。）
- I'd like to show you ...（〜をお見せします。）
- As you can see from this graph ...（このグラフからもわかるように〜。）

5.6　結果の提示例

結果の提示例を以下に示します。

　　実験で受けた圧力による変形の値を決定するのに 15 個の金属に対し水平方向と垂直方向に力を交互に加えてそれらの剛性を測るとする。その実験の結果を次のように記すことができる。

　　個々の金属とそれぞれの条件における剛性の平均値を表 1 に記す。図 1 は圧力の増加による金と銀の変形の様子を示している。金は 40N の圧力で，銀は 55N の圧力で切断されことがわかる。データから，垂直の圧力が変形を引き起こす力が水平の圧力より強いことがわかる。

Chapter 6

Explaining results

The Discussion section of the research paper <u>is where</u> the meaning, significance and relevance of the research results are explored. It explains and compares what has been discovered in the research that is being reported in the paper with the results obtained from previous studies. It thus provides the framework for evaluating and discussing the different questions involved in the investigation and finally makes an argument concerning the best answer to the specific research question the paper is trying to resolve.

is where ⇒ⓖ8.2

6.1 Structure of the Discussion section

The discussion needs <u>to first of all unambiguously state</u> whether the information obtained from the results supports the hypothesis made at the start of the paper. Next, the findings of the particular research being presented in the paper needs to be compared and examined with those that other researchers have obtained. The Discussion section can thus be organized in the following way:

to first of all unambiguously state ⇒ⓖ12.8

(1) Personal explanation of the results
(2) Other possible interpretations
(3) The limitations of the study
(4) Other elements which could have influenced the findings
(5) **Feasible** applications of the results

feasible 実行可能な

The author needs to be sure about whether he has suggested everything that could make his/her findings **invalid**. Consider whether any of the explanations reveal a possible **flaw** (i.e., defect, blunders) in the experimental procedure. Explain whether the analyses contribute or provide new information concerning the research topic investigated. Describe in which cases the new research suggests **shortcomings** or an improvement on the work of others.

invalid 無効な
flaw 欠陥

shortcoming 欠点

The validity of the results in the context of other research needs to be explored. Discuss whether it is possible for the findings to be generalized to other areas.

Future research that may be conducted to give a further explanation of the issues raised because of the information obtained should be described at the end of the Discussion. It should be clarified whether you wish to do these further studies yourself or whether it <u>is to</u> be left to the wider scientific community to pursue.

is to ⇒ⓖ12.7

It is commonly thought that the Discussion section is the most **challenging** <u>one</u> to write. However, having a logical structure to follow as well as a few hours of hard thinking before putting pen to paper should help in making the task much simpler and worthwhile.

challenging 困難だ
one ⇒ⓖ3.2

6.2 Common vocabulary used in the Discussion section

(1) Verbs used to describe research activities

Analyze, assess, compile, determine, develop, discover, evaluate, experiment, explore, find, identify, improve, innovate, investigate, modify, record, search for, study, survey, test, trial

(2) Nouns used to indicate measurements

*Constant, **correlation**, **deviation**, **distribution**, feedback, frequency, **mean**, measurement, median, mode, norm, random, reliability, report, response, sampling, scale, standard,*

correlation 相関
deviation 偏差
distribution 分布
mean 平均値

statistics, *validity*, *variable*, *variance*

statistics 統計

(3) Words that can be used as both nouns and verbs
Experiment, study, test, trial

Examples:
- **Thermodynamics** is the study and application of the thermal energy of systems. (noun)
- One way to study sound waves is to monitor them as they move to the right. (verb)
- Coulomb's law has survived every experimental test. (noun)
- Their purpose was to test Einstein's theory of relativity. (verb)
- In all three trials, the block is pushed through the same distance. (noun)
- The warning system was **extensively** trialed. (verb)
- This was the experiment that led to the discovery of the electron in 1897. (noun)
- The school is trying to experiment with new methods of teaching. (verb)

thermodynamics 熱力学

extensively trialed 広範囲
にわたって試行された

6.3　Useful phrases

When commenting on results, the following words and phrases are frequently used:

- *an increase of about/ roughly/ approximately ...*
- *an initial upward trend is followed by ...*
- *an incease in ...*
- *an upward trend in ...*
- *a peak in ...*
- *a slight / notable / significant* **decrease** *in ...*
- *a slight/ constant/* **marked/ substantial/ increase** *in*

decrease [ˈdiːkriːs]（動詞は
[dɪˈkriːs]）減少
marked 顕著な
substantial かなりの
increase 増加（名詞は i
を，動詞は ea を強く読む）

6.4　Expressing cause and effect

In the Discussion section of a research paper, the reason behind the data obtained and the effect of the data are clarified. Conditional clauses are often used to link purpose and impact to **explicitly iterate** how an event may have caused a particular result. A cause-impact relationship explains the correlation between one event (the cause) making the other event (the impact) occur. One purpose may produce several outcomes.

explicitly 明示的に，こと
ばにして
iterate 繰り返す
cause-impact ⇒ 16.1

> **Difference between *cause* and *reason*:** *Cause* is something that produces an effect, while *reason* is a motive or justification for something.

Table 6.1 Expressing cause

Words or phrases	Example
Because	Because the Earth's rotation is slowing, the length of each day is growing longer.
Since	Since the temperature decreased, the chemical reaction slowed down significantly.
As	As there was no force acting upon the ball, it remained in its initial position.
Due to	That echo was due to a single reflection off the opposite wall of the door.
Owing to	Such measurements became possible owing to the very high precision of modern atomic clocks.
The reason why	This is the reason why a **permanent magnet** has a permanent magnetic field.
For the reason that	The mechanical energy of the system is reduced for the reason that there is considerable **air drag**.

permanent magnet 永久磁
石

air drag 空気抵抗

In that	This equation is equivalent to the other equation <u>in that</u> either can be considered as the defining equation.	<u>in that</u> ⇒ 7.1
Thanks to	The **projectile** underwent a change in linear momentum thanks to the **collision**.	projectile 発射物 collision 衝突
On account of	If a particle is released, it will fall toward the center of Earth, on account of gravity.	

Table 6.2 Expressing effects (or results)

Words or phrases	Example	
So	The experiment was conducted under controlled conditions, so the researchers could accurately observe and analyze the effects of the variable changes.	
Therefore	Therefore, a **diatomic molecule** can only have two **degrees of** rotational **freedom**.	diatomic molecule 二原子分子 degree of freedom 自由度
As a result	As a result of the chemical reaction, a color change occurred in the solution.	
Because of	Because of their great separation, the spheres have no **induced charge**.	induced charge 誘導電荷
Consequently	Consequently, the magnetic field due to the electric **current** is nearly uniform.	current 電流
Thereby	The experimental findings revealed significant changes in the chemical composition, thereby indicating the transformative effects of the reaction.	<u>thereby</u> そのことにより ⇒ 16.9.5
It follows that ...	It follows that the net work is zero for a complete cycle.	It follows that ... その結果…ということになる
Then	If the engine is more efficient than a Carnot engine, then it will **violate** the second law of thermodynamics.	violate（規則などを）破る
In that case	In that case, a true engine can be considered as a Carnot engine.	
Admittedly	Admittedly, this is quite wrong.	admittedly 明らかに ⇒ 5.3
Hence	Hence, an electrically neutral atom has the same number of electrons and protons.	hence このため
Lead to	This leads to the temperature of the **lattice** rising **exponentially**.	lattice 格子 exponentially 指数級数的に
Result in	The **interference** there results in the brightest possible illumination.	interference 干渉
Have as a result	We have, as a result, fully constructive interference.	<u>as a result</u> ⇒ 16.3
End in	You should arrange things now; otherwise this will end in confusion.	
For this reason	<u>It is</u> for this reason that there are often significant differences in road serviceability.	It is for this reason that ... 〜なのはこの理由による ⇒ 強調構文の一例
As a consequence	As a consequence of his theory of special relativity, Einstein demonstrated that mass can be thought of as a type of energy.	
Bring about	**Fermentation** is brought about by some bacteria.	fermentation 発酵

💡 **Vocabulary:** *effects, results, consequences, **implications**, aftermath*

implication 意義（一般的には複数形で用いられる）
aftermath 余波

6.4.1 Difference between *therefore, hence, thus* and *so*

Therefore means *for this reason*, or *because of this or that* - it is related to **deductive reasoning**, it explains WHY this or that is so, or happened.
Hence means *from this/that* and refers to WHERE - position, or point in time; it tells from where or what, or to where or what, something comes, **derives**, or goes. *Hence* usually refers to the future.
Thus means *in this/that way* - it refers to HOW - the manner in which - this or that happens or occurs. *Thus* is usually used to refer to the past. It is frequently used to indicate a conclusion. The primary distinction between *thus* and *so* is that *so* is a conjunction (meaning *and for that*

<u>deductive</u> reasoning 演繹的推論 ⇒ 10

derive 派生する

reason, *and because of that*), whereas *thus* is an adverb (synonymous with *because of that*).

6.4.2　Difference between *consequently* and *accordingly*

Accordingly **denotes** doing something **in accordance with** a method or someone's order. *Consequently* means that something occurs as a result of another action.

denote 示す，意味する
in accordance with ... ～に従って

Exercises

Q1 Complete the following sentences with an appropriate verb from Section 6.2 (1).

i. Tycho Brahe (1546-1601) _____ extensive data without the help of a telescope.

ii. Chinese astronomers believe that the **Crab nebula** is the result of a **supernova** explosion _____ in A.D. 1054.

Crab nebula かに星雲
supernova 超新星

iii. One task is to _____ the various types of energy in the world.

iv. By including **atomic oscillations**, we can _____ the agreement of a **kinetic theory** with experiments.

atomic oscillation 原子振動
kinetic theory 運動論

v. This _____ unification has been going on for centuries.

Q2 Choose the correct phrase in each of the following sentences:

i. Dark regions (result from/result in) fully destructive interference.

ii. The visible wavelengths (result from/result in) fully constructive interference.

iii. Head restraints were built into cars as a (result of/result in) this finding.

Presentation

Q3 Find out about an important test conducted in the automobile industry. Prepare a presentation on the **objectives**, procedure and results of the test.

objective 目標

第6章　結果の説明

　論文の考察セクションでは研究の結果がもつ意味や意義，妥当性が議論されます。本研究で得られた結果を説明し，先行研究で提示されている結果と比較します。それにより，本研究が提起するさまざまな問いを検討する枠組みが得られます。そして，本研究が解決を目指している問いへの最適解を導く議論が可能となります。

6.1　考察セクションの構造

　考察においては，結果から得られた情報が本論文の最初に立てた仮説を立証するのかどうかについてまず第一に明確に述べなければなりません。また，本研究の結果を先行研究の結果と比較検討しなければなりません。したがって，考察セクションは次の順序で構成されます。

(1)　結果の個人的説明
(2)　他に可能な説明，解釈
(3)　本研究の限界
(4)　今回得られた結果をもたらしたかもしれない他の要因の検討
(5)　今回の結果の実行可能な応用

　著者は今回の研究結果を無効としてしまう可能性のあるものをすべて示唆しているかどうかを確認する必要があります。実験の手順の説明に欠陥（すなわち不具合，誤り）の可能性がないか検討してください。分析結果が調査した研究テーマに関する貢献をしたか，あるいは新しい情報を提供しているか説明してください。今回の研究を通して，他の研究の問題（欠陥）が明らかになったり，他の研究の質を高めることになったのであれば，そのことにも触れてください。

　実験結果が他の研究にもたらしうる貢献についても積極的に検討してください。研究・実験結

果を他の領域にも適用できないかどうか議論してください。

　考察セクションの最後に，今回得られた情報（発見）を手がかりに今後の研究でさらに深い分析が行われる可能性があることに言及してください。本研究の前進のため引き続き努力を重ねるのか，それとも他の研究者に託すのか，そのことをはっきり述べてください。

　考察セクションは論文の中でもっとも書きづらい部分だと一般に考えられています。しかし，書き出す前に数時間，論理構成をしっかりと立て集中して考えれば，すっきりとした価値ある考察になります。

6.2　考察セクションでよく用いられる語彙

(1)　**研究活動を表現する動詞**

analyze, assess（評価する），compile（編集する），determine, develop, discover, evaluate, experiment, explore（探索する），find, identify, improve, innovate, investigate（調べる），modify（修正する），record, search for, study, survey（概観する），test, trial（実際に試す）

(2)　**計測を表す名詞**

constant（常数），correlation, deviation, distribution, feedback, frequency（頻度），mean, measurement, median, mode, norm, random, reliability, report, response, sampling, scale, standard, statistics, validity（妥当性），variable（変数），variance（分散）

(3)　**名詞と動詞のどちらでも用いられる語**

test, trial, experiment, study

例：
- Thermodynamics is the **study** and application of the thermal energy of systems.（名詞）（熱力学はシステムの熱エネルギーの研究と応用に関わる研究分野だ。）
- One way to **study** sound waves is to monitor them as they move to the right.（動詞）（音波の研究の一つとして音波の右への移動の追跡がある。）
- Coulomb's law has survived every experimental **test**.（名詞）（クーロンの法則はあらゆる実験により妥当性が示されている。）
- Their purpose was to **test** Einstein's theory of relativity.（動詞）（彼らの目的はアインシュタインの相対性理論の有効性の検証だった。）
- In all three **trials**, the block is pushed through the same distance.（名詞）（これらの3つの試行すべてにおいて，ブロックは等距離を進む。）
- The warning system was extensively **trialed**.（動詞）（警告システムは広範囲に試行された。）
- This was the **experiment** that led to the discovery of the electron in 1897.（名詞）（この実験が1897年の電子の発見につながった。）
- The school is trying to **experiment** with new methods of teaching.（動詞）（その学校は新しい教授法を試行している。）

6.3　役に立つ表現

結果に言及する際，次の語や語句がよく用いられます。

- an increase of about（約）/ roughly（おおよそ）/ approximately（ほぼ）
- an initial upward trend is followed by ...（最初の上昇傾向に続いて～）
- an increase in ...（全体的に～が増加）
- an upward trend in ...（～で上昇傾向）
- a peak in ...（～でピーク）
- a slight（わずかな）/ notable（顕著）/ significant decrease in ...（大幅な減少）
- ... a slight（～わずかな）/ constant（一定の）/ marked / substantial（かなりの）/ increase in（増加）

6.4　因果関係の表現

　考察セクションで，データ（起こった現象）の背後にある理由や引き起こされた結果が明確に示されます。目的と結果（影響）を関連づけて条件節で表現し，「どのような条件下で，ある現象がある結果を引き起こしたかもしれない」ことを明示的に繰り返し述べます。原因 – 結果（影響）の関係（因果関係）でとらえることで，引き起こすものと引き起こされることの相関関係が説明されます。一つの目的が複数の結果を生じさせることもあるのです。

> **cause と reason の違い**：cause は「何らかの結果を引き起こすもの」をいうのに対し，reason は「何かを生じさせた動機」を指します。

表 6.1　原因を示す表現

語と句	例
because	**Because** the Earth's rotation is slowing, the length of each day is growing longer.（地球の自転速度は少しずつ遅くなっているので，1日の長さは長くなっていく。）
since	**Since** the temperature decreased, the chemical reaction slowed down significantly.（温度が下がったので，化学反応は大幅に遅くなった。）
as	**As** there was no force acting upon the ball, it remained in its initial position.（球に力が加えられなかったので，球は動かなかった。）
due to	That echo was **due to** a single reflection off the opposite wall of the door.（反響が生じたのは，ドアの反対側の壁に一度反射したためだ。）
owing to	Such measurements became possible **owing to** the very high precision of modern atomic clocks.（そのような計測が可能になったのは，現代の原子時計の精度が極めて高いことによる。）
the reason why	This is **the reason why** a permanent magnet has a permanent magnetic field.（こういうわけで永久磁石は永久磁場をもっている。）
for the reason that	The mechanical energy of the system is reduced **for the reason that** there is considerable air drag.（このシステムの力学的エネルギーが減少するのは，相当な空気抵抗力が存在するためだ。）
in that	This equation is equivalent to the other equation **in that** either can be considered as the defining equation.（どちらも定義方程式と見なせるという点で，この方程式はもう一つの方程式と同等である。）
thanks to	The projectile underwent a change in linear momentum **thanks to** the collision.（衝突により，発射物の線形運動量に変化が生じた。）
on account of	If a particle is released, it will fall toward the center of Earth, **on account of** gravity.（解き放たれた粒子は重力を受け，地球の中心に向かって行く。）

表 6.2　結果を表す

語と句	例
so	The experiment was conducted under controlled conditions, **so** the researchers could accurately observe and analyze the effects of the variable changes.（実験は制御された条件下で行われたため，研究者は変数の変化の影響を正確に観察し分析することができた。）
therefore	**Therefore**, a diatomic molecule can only have two degrees of rotational freedom.（したがって，2原子分子は2つの回転自由度しかもっていない。）
as a result	**As a result** of the chemical reaction, a color change occurred in the solution.（化学反応の結果，溶液の色が変化した。）
because of	**Because of** their great separation, the spheres have no induced charge.（球体間の間隔が大きいため，球体に誘導電荷は生じない。）
consequently	**Consequently**, the magnetic field due to the electric current is nearly uniform.（したがって，電流による磁場はほぼ一様である。）
thereby（それにより）	The experimental findings revealed significant changes in the chemical composition, **thereby** indicating the transformative effects of the reaction.（実験の結果化学組成に大きな変化が見られ，反応による変質効果が示された。）
it follows that ...（その結果〜ということになる）	**It follows that** the net work is zero for a complete cycle.（その結果，1サイクルの正味仕事量は0になる。）

then	If the engine is more efficient than a Carnot engine, **then** it will violate the second law of thermodynamics.（エンジンが Carnot エンジンより効率のよいものであれば，熱力学第二法則に従わなくなる。）
in that case	**In that case**, a true engine can be considered as a Carnot engine.（その場合，実際のエンジンは Carnot エンジンであると考えられる。）
admittedly（明らかに）	**Admittedly**, this is quite wrong.（明らかに，これはかなり間違っている。）
hence（このため）	**Hence**, an electrically neutral atom has the same number of electrons and protons.（このため，電気的に中性の原子は同数の電子と陽子をもつ。）
leads to	This **leads to** the temperature of the lattice rising exponentially.（この結果，格子の温度は指数関数的に上昇する。）
result in	The interference there **results in** the brightest possible illumination.（そこでの干渉により，可能な限り明るい照明が得られる。）
have as a result	We **have, as a result**, fully constructive interference.（その結果，期待した干渉が得られる。）
end in	You should arrange things now; otherwise this will **end in** confusion.（混乱が生じないよう，今手配しておかなければならない）
for this reason	It is **for this reason** that there are often significant differences in road serviceability.（このため，道路の整備性に大きな違いが生じることがよくある。）
as a consequence	**As a consequence** of his theory of special relativity, Einstein demonstrated that mass can be thought of as a type of energy.（特殊相対性理論により質量はエネルギーの一形態となることを，アインシュタインは示した。）
bring about	Fermentation is **brought about** by some bacteria.（発酵はバクテリアにより生じる。）

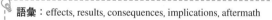

語彙：effects, results, consequences, implications, aftermath

6.4.1　therefore, hence, thus および so の違い

therefore は for this reason や because of this or that の意味をもち，演繹的推論（deductive reasoning）と関係します。「なぜ」そうなのか，「なぜ」そうなったのかを示します。

hence は from this/that を意味し，「どこで」を指します。場所やある時間のことです。「どこから」とか「何から」，あるいは「どこへ」，「何へ」あるものがやってきたり，派生したり，出て行くのかを示します。hence は未来を指します。

thus は in this/that way を意味し，「どうやって」に言及します。どうやって出来事が起きるのかを示します。thus は通常過去の出来事を述べます。結論を述べる際に多用されます。

thus と so の根本的な違いは，so は接続詞（and for that reason, and because of that の意味）であるに対し，thus は副詞（because of that と同義）です。

6.4.2　consequently と accordingly の違い

accordingly はある方法や人の指示に従って実行することを述べます。consequently は，あることが別の行為の結果として起こるときに使われます。

Chapter 7

Analyzing graphs

Graphs are a visual way of presenting **statistics**, in particular experimental data. Analyzing graphs is a useful way for determining the general trend of the data, relating the results of an experiment to the initial hypothesis, and for formulating hypotheses for future experiments.

statistics 統計

7.1　Types of graphs

The three most important types of graphs are:

(1) Line graph - used to show changes over short and long time periods. They are quite **informative** in allowing the reader to visualize trends.

(2) Bar graph - uses both horizontal and vertical bars to show comparisons amongst different types of data.

(3) Pie chart - helps to visualize the relative importance of several properties of a **variable**.

line graph 折れ線グラフ
informative 的確な情報を与える，有益な
bar graph 棒グラフ

pie chart 円グラフ
variable 変数

When there are minor changes, line graphs are preferable to bar graphs. Line graphs can also be used to compare changes in more than one group over the same time period.

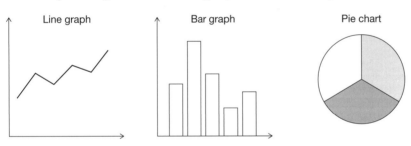

Line graph　　　Bar graph　　　Pie chart

7.2　Analyzing information from graphs

When analyzing information in a line or bar graph, the time periods should be noted as well as the amount increased or decreased during that interval.

First, the **axes** of the graphs should be studied in order to determine what sort of information is being represented. The *x*-axis represents the **independent variable**, or that which can be changed. The *y*-axis shows the **dependent variable**, or that which relies upon the independent variable. Then the general trend of the graph should be determined. Experimental points that do not seem to follow the general trend need to be examined closely. Not all sets of experimental data show a trend. If there is one bar, dot or part of the line that is **out of place**, then it may **not** have **that much** of an important effect on the complete result. Finally, the graph should be used to make predictions about future experimental data.

a line or bar graph ⇒ G 2.5.2

axes 軸（単数形は axis）
is being represented ⇒ G 10.2
independent variable 独立変数
that which ⇒ G 8.4
dependent variable 従属変数
do not ⇒ G 16.10
out of place 場違いの，重要でない
not ... that much それほど～でない

7.3　Useful phrases to refer to graphs

The following phrases can be used to refer to graphs during a presentation or in a research report.

- The ... represents ...
- Take a look at this.
- The graph illustrates
- On this graph, ...
- I'd like to draw your attention to ...
- Let's have a look at this.
- If you look closely
- I'd like you to look at this.
- Here we can see ...
- As you can see, ...

7.4 Commenting on graphs

When writing the Results section, graphs are used to illustrate points in the textual content. **Instead of** simply telling readers what they are able to already see, it is much more beneficial to write about the special features that may be **inferred** from the graph. Otherwise, the reader is forced to make his or her own interpretations (which may not be the correct interpretations). It is better to save the readers from exerting any mental effort and at the same time influencing them to look **favorably** on the interpretation of the experimental data that is **consistent** with your own **deductions**.

instead of … 〜ではなく
infer 推論する

favorably 好意的に ⇒Ⓖ5.4
consistent 一致している，
首尾一貫した ⇒Ⓦ11
deduction 推論 ⇒Ⓦ10
conciseness 簡潔

Lack of conciseness

Eliminate phrases such as "as may be seen." and "we can see." Instead, include the figure or table reference number in brackets at the conclusion of the sentence. In the case where readers are being referred to a graph, there is no need to use words like *graphically* or *schematically*. Also, many different synonyms should not be applied to describe what kind of graph it is or to explain what it suggests. If possible, use active verbs – "this figure indicates *x*.", instead of "*x* is indicated in this figure.". In the text discussing the graph, facts should not be **duplicated** which can already be clearly understood by studying the graph. Only the highlights need to be expressed unambiguously.

duplicate 繰り返す

As opposed to just repeating the statistics that can be seen on the graph, the data should be interpreted by comparing the outcomes. Only the key results or trends that the graph conveys should be pointed out.

as opposed to … 〜するのではなく

Another common mistake is to copy **word for word** the **caption / legend** on the graphs and tables when referring to these in the text. Legends for graphs should be concise and quite detailed, enabling readers to understand the meaning of the graph without having to read the accompanying text. Therefore, it is essential that the legends should be carefully constructed as some readers may find it most effective to just look at the graphs and tables without reading the main paper itself.

word for word 一語一語
caption キャプション
legend 凡例

Chapter 7

Exercises

Writing

Q1 Consider a bicycle trip through a hilly country and a walk in the park. Draw one graph showing distance versus time for both of these two situations. Analyze the graph and compare the speeds for the two cases. The data should be fictional but reasonable. After 30 minutes, what predictions can be made for the future?

Q2 In the case of **maritime disasters**, one hypothesis is that crew members have a survival advantage over passengers. After doing research into this subject, write a short report using two or more graphs proving or disproving the hypothesis.

maritime disaster 海難事故

using ⇒Ⓖ11.2

第7章　グラフの分析

　グラフは統計，特に実験データを目に見える形で示します。グラフを適切に分析すれば，データの特徴をつかみ，結果を最初に立てた仮説と関係づけることができ，次の実験の仮説を立てることができます。

7.1　グラフのタイプ

　グラフには以下の3つの最も重要なタイプがあります。

(1)　**折れ線グラフ**：短期または長期の時間経過での変化を示す。読者がトレンドを視覚化するのに，折れ線グラフはかなり有益な情報である
(2)　**棒グラフ**：縦横の棒を使うことで，異なる種類のデータの比較ができる
(3)　**円グラフ**：ある変数のもつ，いくつかの性質の相対的な重要性が，一目で見てとれる

　変化が大きくない場合には，棒グラフより折れ線グラフが便利です。折れ線グラフは，同じ時間経過の中で異なるグループに見られる変化を比較するのにも用いられます。

7.2　グラフを用いた情報の分析

　線グラフまたは棒グラフで情報を分析する際，期間とその間の増減の量を記さなければなりません。
　まず，グラフの縦横の軸により情報の種類を確認します。x軸は独立変数（自由に変更を加えることのできるもの）を，y軸は従属変数（独立変数に応じて変化するもの）を表します。次に，グラフの増減の傾向を決定します。傾向とは異なる箇所があれば，それについての考察が必要です。すべての実験データが特定の傾向を示すとは限りません。グラフの線や棒の中の1ヵ所が傾向からはずれていても，それが全体の結果にそれほど重要な影響を与えないかもしれません。最後に，グラフから今後の実験データの予測を導く必要があります。

7.3　グラフを説明するのに用いられる表現

　論文や口頭発表では，以下の表現を用いてグラフを説明します。
　・The ... represents ...（〜は〜を表します。）
　・Take a look at this.（これを見てください。）
　・The graph illustrates（グラフは次のことを示しています。）
　・On this graph, ...（このグラフでは，〜）
　・I'd like to draw your attention to ...（注目していただきたいのは〜）
　・Let's have a look at this.（これを見てみましょう。）
　・If you look closely（よく見ると）
　・I'd like you to look at this.（これを見ていただきたい。）
　・Here we can see ...（ここでわかるのは〜）
　・As you can see, ...（ご覧のとおり，〜）

7.4　グラフについてのコメント

　結果を書く際には，グラフを用いて本文中の重要なポイントを示します。グラフを見れば自明のことを述べるのではなく，グラフから読みとれることに言及してください。それを怠ると，読者は自ら解釈をしなければならなくなります。そしてその解釈は誤っているかもしれないのです。読者に無用の努力を強いることを避けるだけでなく，著者の推論と合致するグラフの解釈を示して，読者の理解を得るようにしてください。

簡潔に書く

　"as may be seen."（見ての通り）や"we can see."（見て取ることができます）などのフレーズを削除し，代わりに，文の終わりに図または表の参照番号を括弧内に含めます。グラフに言及する際，graphicallyやschematicallyといった表現は不要です。また，グラフを説明したり，グラフから読みとれることを述べるとき，類義語を並べてはいけません。できるだけ能動態を使いましょう。たとえば，"x is indicated in this figure."ではなく"this figure indicates x."を使います。
　グラフに見られる統計をくり返すのではなく，種々の結果を比較することでデータを解釈してください。グラフの示している重要な結果や傾向のみを指摘してください。
　もう一つのよくみられる間違いは，テキストで述べる際に，グラフや表のキャプション／凡例をそのまま写すことです。本文のテキストを読まなくてもグラフの意味を読者が理解できるよう，グラフの凡例は簡潔かつ詳細である必要があります。読者の中には，論文自体を読まずにグラフや表を見るのが最も効果的であると思っている人もいます。したがって凡例を注意深く作成することが不可欠です。

Chapter 8
Summarizing

At the end of a report or research paper, the Conclusion section provides a summary to give the reader a brief reminder of the main ideas and results of the research project, while restating the main research problem. It is important to remember that the Conclusion is the final idea **left with the reader** at the end of a research paper.

left with the reader 読者に
届けられる

8.1 Difference between a summary and the Conclusion section

Although a summary of what was learned from the research should be included in the Conclusion part of a research report, this summary should be kept to a minimum because the focus of the conclusion should be on the consequences, assessments, and **insight**s of the research. Often, it is stated in the conclusion which hypotheses the author considers to be supported by the most compelling evidence. While the Conclusion section for some journals is referred to as a Summary, it is **imperative** to remember that a Conclusion is not a summary only.

insight 洞察

imperative … ～しなければ
ならない

8.2 Stages of a summary

The major points of an entire paper can be summarized in the following order:

: ⇒ 16.2.1

    (1) the main goal of the study and the research problem(s) under investigation;

; ⇒ 16.2.1

    (2) the fundamental design of the study;
    (3) major findings or patterns identified through analysis;
    (4) a brief explanation of the author's interpretations and conclusions.

8.3 Methods of summarizing

Readers first get a general idea of a research paper by reading the Abstract and Conclusion sections, which basically provide a summary of the whole project. The Conclusion part helps a reader understand if the goal summarized in the abstract has been reached. All the key points that have been made throughout the body of the paper, i.e., the things that have been stated to prove the initial hypothesis need to be summarized.

To write a summary, understanding the topic, the gist, or the larger conceptual framework of a paper, book, article, paragraph, or a passage is required, and the first step is to read the subject under discussion in an intelligent way. Just by reading the first and last paragraph of an article or the first and last sentence of a paragraph, you can usually get a pretty good idea of what the **extract is about**.

The first and last sentences of any extract are therefore **crucial** in understanding the main idea.

extract [ekstrækt] 抜 粋,
要旨 ⇒ 12
is about 全体の内容がどの
ようなものであるか
crucial 極めて重要な
in any event いずれにせよ

all in all 全体として見れば

> **Words used in a summary: In any event**, as mentioned, therefore, as has been said, in conclusion, in sum, as shown, on the whole, consequently, in other words, in summary, in short, hence, in brief, that is, finally, in the final analysis, to conclude, as has been noted, altogether, thus, **all in all**, in a word, summing up

Exercises

Writing

Q1 Write a 1-sentence-long summary of the following paragraph:

Climate change has **detrimental** effects on livelihoods, habitats <u>and</u> infrastructure. So, green strategies are necessary, but it is a question of which ones are the most effective. Natural gas is a fossil fuel that burns relatively cleanly. However, there are some environmental and safety concerns with the production and use of natural gas. The environmental **credentials** of natural gas are harmed by **methane** leaks. Renewable energy is generated by **inexhaustible** fuel sources that restore themselves over a short period of time and do not **diminish**. Solar energy is an <u>easy-to-use</u> huge renewable energy source. But large solar power plants, like any other type of power plant, can have an impact on the environment. Hydropower is <u>non-polluting</u> but <u>does also have</u> environmental impacts. When the water of a dam is released, it can have negative effet on downstream plants and animals. Nuclear power is now viewed as a potential, **albeit** possibly dangerous, source of green power. However, a terrorist attack on a nuclear power plant is a hypothetical **catastrophic eventuality**.

detrimental 有害な
<u>and</u> ⇒ⓖ16.2.2

credentials 信頼
methane [meθeɪn] メタン
inexhaustible [ˌɪnɪɡˈzɔːstəbl] 無尽蔵の
diminish 減少する
<u>easy-to-use</u> ⇒ⓖ16.1
<u>non-polluting</u> ⇒ⓖ16.1
<u>does have</u> ⇒ⓖ16.6
albeit … [ˌɔːlˈbiːɪt] 〜だが
catastrophic eventuality 壊滅的な事態

Q2 Choose an article from the journal *Nature* and write a summary of approximately 50 words explaining the gist of the article.

第 8 章　要約

　レポートや学術論文の最後に置かれる結論セクションで，研究プロジェクトの主要な考えと結果の要約を読者に示すと同時に，主要な研究課題を再度述べます。結論は研究論文の本旨を読者に届けられる最後の機会であることを忘れずにいてください。

8.1　要約と結論の違い

　論文の結論セクションでは研究で明らかになったことを要約しますが，要約は短く書かなければなりません。結論の中心は研究の結果や評価，今後の見通しにあるからです。多くの場合，どの仮説が最も説得力のある証拠によって裏付けられていると著者が考えているかが結論で述べられます。結論＝要約ととる学会誌もありますが，結論は要約にとどまらないことに留意しなければなりません。

8.2　要約セクションに記されるもの

　論文の要約で主に記されるのは，以下の 4 つです。この順序で展開されます。

(1)　研究の主な目的と調査中の研究課題
(2)　研究の基本的方針
(3)　分析によって明らかになった主な発見またはパターン
(4)　著者の解釈と結論の簡潔な説明

8.3　要約の方法

　プロジェクト全体の要約を記したアブストラクトと結論のセクションを読むことで，読者は研究論文の全体像を把握します。結論を読むことで，アブストラクトで記された研究の目標が達成されたかどうかを知ることができます。それまでに述べてきた重要なポイント，すなわち最初の仮説を証明するために述べてきたことをすべて要約しなければなりません。

　要約を書くには，論文や書籍，記事，パラグラフ，文章などのトピックや要旨，またはより大きな概念的枠組みの理解が求められます。その第一歩は，議論されている対象を論理的に読むことです。たいていの場合，論文の最初と最後のパラグラフ，あるいはパラグラフの最初と最後の文を読むだけで，その抜粋が何について書かれているのかをかなり理解できます。

　したがって主要な考えを理解するのに，どのような抜粋であってもその最初と最後は極めて重要なのです。

要約の際用いられる語：in any event, as mentioned（前述のように）, therefore（したがって）, as has been said（前に言ったように）, in conclusion（結論として）, in sum（要するに）, as shown（示したように）, on the whole（全体として）, consequently（結果として）, in other words（言い換えると）, in summary（要約すると）, in short（要するに）, hence（それゆえ）, in brief（簡潔に言えば）, that is（つまり）, finally（最終的に）, in the final analysis（最終的な分析において）, to conclude（結論として）, as has been noted（既に記したように）, altogether（要するに）, thus（このように）, all in all, in a word（一言で言えば）, summing up（要約すると）

Chapter 9

Recommending actions

In the Conclusion section of a research paper, recommendations may sometimes be included. When writing the recommendation part, <u>one</u> needs to state which other actions should be taken as a result of the research. The recommended activities should refer to future work for the author and/or the entire research community. <u>**Also included are**</u> suggestions for improvements, possibly in relation to the limitations observed in that specific research effort.

one ⇒ G 3.1

also included are … 他に～も含まれる ⇒ G 16.4

9.1 Language expressing recommendations

(1) *Should/ought to* is used to show that you are recommending or suggesting an action:

Example:
· You still should be able to use a conversion table.

(2) *Should + passive* is used in the recommendation part of a report:

Example:
· Numbers up to three significant figures should be entered in the column.

(3) If *propose/recommend/suggest* is used, the construction is:

<u>to recommend (suggest, propose)</u> + <u>that</u> + <u>someone or something</u> + <u>should (optional)</u> + <u>infinitive (without to)</u> + something

Examples:
· We <u>propose</u> <u>that</u> the hour <u>(should)</u> <u>be</u> redefined to provide a 10-hour day.
· We <u>propose</u> <u>that</u> the scientific committee <u>(should)</u> <u>redefine</u> the hour to provide a 10-hour day.

The form of the second verb does not change in the second clause – in fact it is an infinitive form. This means that no third person *–s* is required in correct English, hence "*We propose that the scientific committee <u>redefines</u> the hour to provide a 10-hour day.*" is <u>incorrect</u>. The first sentence uses the passive infinitive (*be*) + past participle (*redefined*).

9.2 Phrases to use when making a suggestion

When speaking, the emphasis should be on achieving the goal rather than **advocating** for a particular approach to achieving it. Your words should make **alternative** interpretations and actions seem **plausible** to the audience.

advocate 主張する
alternative 別の
plausible 妥当と思われる

(1) Suggestions can be made <u>using</u> a variety of forms:

· Try ...
· Why don't we do ... ?
· You could ...
· I would suggest that we do ...

using ⇒ G 11.2

(2) When the suggestion is for yourself and others in a group, the following forms may be used:

· Let's try doing ...

· Let's do ...
· Why don't we ...
· What / How about doing ... ?
· We could do ...

(3) Questions asking for suggestions:

· What should I/we do?
· What do you suggest I/we do?
· What do you think we should do?

At a conference or seminar, the audience should be told that you will welcome speaking to anyone who can suggest ways of improving your line of research. During the last part of a presentation, it is best to **dwell on** how your future work will **progress,** especially how you intend to **rectify the problems** you have encountered.

dwell on 考える
progress [prəˈgres] 進展する
especially ⇒ⓖ5.5
rectify the problem 問題を修正する，問題に寄り添い適切に対処する（correctにする）

9.3　Ways of making suggestions

The words *propose, suggest* and *recommend* are usually used when recommending actions.

(1) Propose + **that** expresses specifically who will carry out the proposed action
(2) Propose + -ing does not **specify** who will carry out the action and is often used when the person who will act presents the proposal

specify 明示する

Example:
· We propose planning new experiments.

Difference between suggest, propose, and recommend

The words *propose* and *proposal* in a research or a work context **signifies** something formal or official which requires a formal acceptance or rejection.
Suggest and *recommend* are commonly used to signify something that is not very formal. *Suggest* is used when talking about giving someone an idea in general, and *recommend* is used when telling someone that a certain choice or option is the best one.

and ⇒ⓖ16.2.2
signify 意味する
rejection ⇒Ⓦ13

Examples:
· We propose to accept the other project.
· The position of the iron **filings** suggests the existence of magnetic field lines.
· These are the values recommended by the agency.

filing やすり

Exercises

Writing

Q1 Choose the correct word:

i. The **metric system** was *proposed/recommended/suggested* in 1792.

metric system メートル法

ii. The experiment *proposed/recommended/suggested* by this sample problem has been done.

iii. In 1956, Frank Lloyd Wright *proposed/recommended/suggested* the construction of a mile-high building in Chicago.

a mile-high ⇒ⓖ16.1

Q2 Make some recommendations for improvements to your university. Use *should* with the **passive infinitive**.

passive infinitive 受動不定詞

Chapter 9

第 9 章　行動の推奨

　研究論文の結論セクションに推奨を書くことがあります。推奨パートを書く際，研究の結果として，今後他にとるべきアプローチを述べる必要があります。推奨は，著者の，かつ（あるいは）研究分野全体の今後に言及すべきです。また，今回の特定の研究の問題点にからめて，よりよい研究についての示唆も述べられます。

9.1　推奨の表現

(1)　should/ought to を用いて推奨または示唆であることを表します。

　　例：
　　・You still **should** be able to use a conversion table.（今でも換算表を使えるのが望ましい。）

(2)　レポートの推奨部分では should + passive も用いられます。

　　例：
　　・Numbers up to three significant figures **should be entered** in the column.（コラムに，重要な図を最大 3 つ入れた方がよい。）

(3)　propose/recommend/suggest は次の形で使われます。

　　to **recommend**（**suggest**, **propose**）+ that + someone or something + should（あってもなくてもよい）+ 不定詞（to は用いない）+ something

　　例：
　　・We **propose that** the hour (should) be redefined to provide a 10-hour day.（時間の概念を再定義し，1 日は 10 時間からなると仮定してはどうか。）
　　・We **propose that** the scientific committee（should）redefine the hour to provide a 10-hour day.（科学委員会は時間の概念を再定義し，1 日 10 時間としてはどうか。）

　2 番目の節の動詞の形は変わらず，不定詞の形になっています。つまり，正しい英語では三人称の −s は不要であり，したがって "We propose that the scientific committee <u>redefines</u> the hour to provide a 10-hour day." は誤りです。第 1 文は受動態不定詞（be）＋過去分詞（redefined）を使っています。

9.2　示唆の表現

　話すときは，目標達成への特定のアプローチを主張せず，目標達成の重要性に重点を置くべきです。別の解釈や行動がより妥当性があると聴き手に思わせるように話すべきです。

(1)　示唆はさまざまな形で行うことができる：
　　・Try …
　　・Why don't we do … ?
　　・You could …（〜できるかもしれない）
　　・I would suggest that we do …

(2)　示唆が発表者と共同研究者の両方に向けられている場合，次のように言う：
　　・Let's try doing …
　　・Let's do …
　　・Why don't we …
　　・What / How about doing … ?
　　・We could do …

(3)　示唆を求める質問：
　　・What should I/we do?
　　・What do you suggest I/we do?
　　・What do you think we should do?

　学会やセミナーでは聴き手に対し，発表した研究の質を深める示唆を歓迎する旨を伝えてください。発表の最後に，今後の研究の方向性，特に，現時点での問題点へのあなたの取り組みに言及してください。

9.3　示唆の仕方

　示唆や推奨をするには，propose, suggest, recommend という動詞が用いられます。

(1)　**propose + that** は提案されたことがらを実行するのは誰なのかを明示します。

(2) **propose** + **-ing** は提案されたことがらを実行するのは誰なのかを示しませんが，この表現は，実行するのが発表者である場合に多く使われます。

例：
　　・We **propose** planning new experiments.（新しい実験を考えてみてはどうか。）

suggest, propose, recommend の違い

　研究や仕事の文脈で用いる propose や proposal という単語は，正式な受け入れや拒否を必要とする正式なもの，公式なものを意味します。それに対し suggest と recommend はそれほどフォーマルではありません。一般的に suggest は誰かにアイデアを与えるときに使われ，recommend は特定の選択肢やオプションがベストであることを伝えるときに使われます。

例：
・We **propose** to accept the other object.（もう一方のやり方がよい。）
・The position of the iron filings **suggest** the existence of magnetic field lines.（鉄のやすり屑は磁力線の存在を示唆している。）
・These are the values **recommended** by the agency.（これらはその組織が推奨する値だ。）

Chapter 9

Chapter 10
Writing an abstract

In a paper the abstract comes at the beginning, but you should write at the end. Articles start with a summary paragraph called an abstract, <u>**ideally**</u> not more than 200 words. It contains no references and serves as a brief introduction to the topic and a non-technical **synopsis** of the main results and their **implications**. The abstract needs to be clearly written and readily comprehensible to everyone. It can be read as **a stand-alone text** without any **footnotes** - it should not be considered as **part** of a paper. Many readers will only read the abstract of the manuscript. So **basically**, it is the abstract which helps the reader determine whether they should read the entire article or not.

ideally 理想的には，できれば ⇒ G5.4
synopsis あらすじ
implication 意義（一般的には複数形で用いられる）
stand–alone text 独立した（他のテキストの参照を必要としない）テキスト
footnote 脚注
<u>**basically**</u> 基本的には ⇒ G5.4

10.1　Structure of an abstract

An abstract is a <u>150- to 250-word</u> paragraph that gives readers a short evaluation of your research and its significance. It should summarize the main idea of the paper, <u>emphasizing</u> essential elements; it should also state any implications or applications of the findings presented. The information provided within the abstract needs to be sufficient to assist other researchers in determining whether the article is **relevant** to his or her interests or not. The easiest way of writing an abstract is to imitate the structure of the larger manuscript — it can be considered as a miniature model of your complete research paper. The same sections of the main paper can thus be **replicate**d in the abstract: Introduction, Method, Results, Discussion, and Conclusion. It is required to begin with a brief explanation of the study problem, followed by a short outline of the research methodologies and results, and finally a concise discussion of the primary implications of the findings.

<u>150- to 250-word</u> ⇒ G16.1
<u>emphasizing</u> ⇒ G11.2
; ⇒ G16.2.1

relevant 関連のある

replicate そのまま使う

> **Background information in an abstract:** The background information section of an abstract assists the reader in determining whether the content offered about a research subject is worth reading and promotes trust in the research and conclusions.

10.2　Key skills of writing an abstract

The abstract should provide a concise yet **comprehensive** summary of the contents of the paper.
The abstract should include as many key words as possible related to the content of the text<u>:</u> these will make it easier to find the abstract in computer searches. While search engines ignore **capitalization rules** by default, hyphenation may be problematic; therefore, spelling should be carefully addressed.
An abstract is not a review, <u>nor does it comment</u> on the work performed. It ought to outline the most essential factors of the study while providing only a limited amount of detail on the background, **methodology** and **consequences**.
Each sentence should be written with consideration of how to make the maximum impact. The abstract should only concentrate on stating four or five essential points, principles, or findings. It also needs to be **objective** and correct.

comprehensive 包括的な

<u>:</u> ⇒ G15.2 (2)

capitalization rule 大文字の使用に関する決まり
; ⇒ G16.2.1
<u>nor does it comment</u> ⇒ G16.4
methodology 研究手法
consequence 実験の結果もたらされるもの（たとえば社会的影響）
objective（形容詞）客観的な

10.3   Abstract writing style

A **well-written** abstract <u>serves several purposes</u>: it allows readers to quickly **grasp** the **gist** or essence of the paper or article. The function of an abstract is to explain what has been done, <u>not</u> to **assess** or **expose** the limitations or strengths of the paper. An effective abstract is not a collection of manuscript sentences; instead it is a reworded gist.

The abstract should be written **at the very last**, after the rest of the text has been finished. **Breaks** should not be used inside the abstract.

There may frequently be a strict word restriction, so make sure to check the requirements of the university or journal. In a **dissertation** or **thesis**, the abstract has to be written on a separate page, after the title page and **acknowledgements** but before the **table of contents**.

10.3.1   Beginning an abstract

The first sentence of an abstract introduces the research problem or outlines a particularly interesting field in current research. Using different phrases, it should seek to establish the relevance or importance of the paper to the wider research area and confirm the importance of the research conducted. When beginning the abstract, you should thus ask yourself **what need it meets**, what gap in pioneering research it seeks to fill or what problem it addresses. In many cases, the use of a first-person pronoun, like 'we', is **discouraged**. If you have been the person (or part of a group) who carried out the main research, usage of the first person is sometimes acceptable.

10.3.2   Abstract tenses

When writing an abstract for a research article, several **tenses** can be used. The tense that you should use depends on the subject of the sentence. It is a well-known rule that <u>any statement of fact</u> ought to be written <u>using</u> the present tense. The past tense should be used when writing about previous research. Usually, the simple present tense is used when stating facts and data and also when explaining the consequences of your results. The past tense should be used for describing your methodology and particular findings. Either one of those tenses can be used when writing about the reason behind the research. The verb tenses in an abstract should **correspond to** the tenses used in the main paper.

10.3.3   Keywords

Keywords are very important elements of an abstract; they help researchers discover content, <u>especially</u> through a search engine. A collection of the keywords of an abstract should be able to almost describe the entire work. <u>Interestingly</u>, three factors — the title, abstract and keywords — may hold the **secret** to publication success.

10.4   Conference abstract

Conference abstracts are presented to conference organizers, who usually have many more submissions than presentation slots, so the conference abstract needs to be carefully written to **enhance** the probability of <u>it being accepted</u>. Every conference has rules about <u>the writing of</u> the abstract on its website; these should **be** carefully **adhered to**, especially the style and spelling should be exactly as the organizers wish. There is a **conventional** method of writing a conference abstract: the stages of such an abstract are generally 1. subject matter, 2. motivation, 3. research problem, 4. method, 5. results, 6. consequences and conclusions. The abstract is also intended for the study of conference **attendees** since abstracts **are** usually **indexed** in conference programs. Therefore, an effort should be made so that the abstract is clear and attractive to **facilitate the audience feeling an interest** in attending the main oral presentation on the conference day. Conference abstracts are usually longer than an abstract of a re-

---

well-written よくまとめられた
<u>serves several purposes</u> ⇒G16.5
grasp 把握する
gist 要旨
<u>not to assess or</u> ⇒G16.7
assess 評価する
expose 浮きぼりにする
; ⇒G16.2.1
at the very last 最後の最後に
break 1つの文章を複数のパラグラフにわけること
dissertation 論文
thesis 論文
acknowledgements 謝辞
table of contents 目次
what need it meet どのような必要用件に応えているか ⇒G16.5
discourage くじけさせる，薦めない（encourage の反意語）

tense 時制
<u>any statement of fact</u> ⇒G3.6, 3.8
<u>using</u> ⇒G11.2

correspond to 一致する

especially ⇒G5.5
interestingly ⇒G5.3
secret 秘訣

enhance 高める
<u>it being accepted</u> ⇒G13
<u>the writing of</u> the -ing of の形をとる。（例：The Making of the English Bible）
be ... adhered to 遵守される ⇒W17
conventional 慣習的な ⇒W14
attendee [əˌtenˈdiː] 出席者
be ... indexed 索引に掲載される
facilitate the audience feeling an interest 聴き手に関心を呼び起こす

search paper, <u>divided</u> into paragraphs and contain one or more figures.　　　　　divided ⇒ G11.2

Exercises

Q1 What are some of the typical characteristics of poor abstracts?

Writing

Q2 Read an article on how ice can cause an airplane to crash. Then write an abstract for the article in approximately 150-200 words. Explain it in two separate paragraphs:

　i. one technical

　ii. other non-technical

第10章　アブストラクトを書く

　論文でアブストラクトは最初に置かれますが，書くのは最後です。論文の最初に置かれるのはアブストラクトと呼ばれる要約のパラグラフです。できれば200字を越えないように書いてください。それは参考文献を含まず，研究トピックについて簡潔に紹介し，主な結果とその意味について一般的な用語で概略を記したものです。アブストラクトは誰でも一読して理解できるよう明確に書かれていなければなりません。それは論文の一部ではなく自立したパラグラフであり，脚注をもちません。多くの読者はアブストラクトしか読みません。基本的にアブストラクトに魅力を感じて初めて論文を読んでもらえるのです。

10.1　アブストラクトの構造

　アブストラクトは150〜250語からなり，研究の内容と重要性を読者に示します。アブストラクトは論文の主要な考えを要約し，最も重要な点を強調します。また提示した研究成果のもつ意義と応用も述べるべきです。アブストラクトを読んだだけで他の研究者が彼らの興味，関心を引くものであるかどうかを判断できるのでなければなりません。アブストラクトを書く最も簡単な方法は，大きな原稿の構成を真似ることです。それは研究論文全体の縮図と見なすことができます。そのため，本論文と同じセクション（序論，方法，結果，考察，結論）をアブストラクトでも再現することができます。研究を行っている問題の簡単な説明から始まり，研究方法と結果の簡単な概説が続き，最後に研究結果のもつ主要な意味を簡潔に論じることが求められます。

> 💡 **アブストラクトにおける背景情報**：アブストラクトの背景情報セクションは，研究テーマについて提供された内容が読むに値するかどうかを読者が判断する助けとなり，研究や結論に対する信頼を高めます。

10.2　アブストラクト作成に際しての主要なスキル

　アブストラクトは，本論文の内容を簡潔に，しかし，全体的に述べていなければなりません。

　アブストラクトには，論文の内容に関連するできるだけ多くのキーワードを含める必要があります。それにより，コンピュータ検索でそのアブストラクトを見つけやすくなります。検索エンジンはデフォルトでキャピタライゼーションルールを無視しますが，ハイフンは問題になる可能性があります。したがって，スペルは慎重に対処する必要があります。

　アブストラクトはレビューでもなければコメントでもありません。それは，研究の最も重要な部分を概説する一方で，研究の背景や研究手法，実験結果（およびそれがもたらすもの）を簡潔に書きます。

　各文は読者に印象的に映るよう書かれなければなりません。アブストラクトは4つか5つの重要な点や原理，研究結果のみを述べなければなりません。また，客観的で正確でなければなりません。

10.3　アブストラクトのスタイル

　よくまとめられたアブストラクトにはいくつかの目的があり，その一つは，読者が論文の要旨

やエッセンスを一読で理解できることです。アブストラクトの役割は行った研究を説明することであり，執筆した論文のすぐれたところと限界を評価したり浮きぼりにしたりすることではありません。質の高いアブストラクトは論文本体の文の寄せ集めではなく，再度書き換えられた要旨です。

　アブストラクトは論文本体をすべて書き終えた後，最後に執筆します。アブストラクトは1パラグラフにおさめなくてはいけません。

　字数制限は厳格に守らねばなりません。大学やジャーナルで定めた制限字数を確認しておいてください。学位論文では，アブストラクトはタイトルと謝辞のページの次，目次の前のページに置かれます。

10.3.1　アブストラクトの出だし

　アブストラクトの第一文は，研究課題を紹介したり現在の研究のとくにホットな分野を概説したりします。いろいろな表現を用いて，より広い研究領域との関連性や論文の重要性を示し，行った研究の重要性を確認すべきです。アブストラクトを書き始める前に，本論文がどのようなニーズを満たしているのか，先端的研究で未解決の問題を解決しようとしているのか，あるいはどのような問題に取り組んでいるのかを確認してください。

　一人称の使用はふつう避けられます。ただし，単独の研究であるか，研究で主要な役割を果たしたのであれば，一人称が受け入れられることもあります。

10.3.2　アブストラクトにおける時制

　論文のアブストラクトではいくつかの時制を使うことができます。各文の主語により時制は決定されます。まず知っておくべきは，事実を述べるのには現在形を用いる，という規則です。先行研究への言及では過去形を用います。事実やデータを述べたり，研究結果を説明する際には，現在形を使うのがふつうです。手法や得られた結果は過去形で報告します。研究の背景は現在形と過去形のどちらで述べてもかまいません。アブストラクトで使う時制は，論文本体で使う時制と同一でなければなりません。

10.3.3　キーワード

　キーワードはアブストラクトのとても重要な部分を占めます。他の研究者は，キーワードを手がかりに検索エンジンでこのアブストラクトにたどり着くのです。並べられたキーワードを眺めれば論文本体の全体像がつかめるように，キーワードを選ばなければなりません。興味深いことに，タイトルとアブストラクト，そして，キーワードは，論文が成功する鍵を握っています。

10.4　学会に提出するアブストラクト

　conference abstract は学会主催者に提出するアブストラクトです。実際に発表できる枠よりはるかに多くのアブストラクトが提出されます。そのため，受理されるようアブストラクトの執筆には最大限の努力が求められます。どの学会もアブストラクトの書き方をウェブサイト上で明確に指示していますので，熟読し，指示に従ってください。特に，スタイルと綴りには注意が必要です。学会に提出するアブストラクトは通常次の順序で書かれます。1. 向き合っているトピック 2. 動機 3. 研究課題 4. 方法 5. 結果 6. 結果と結論です。学会への参加者のためにもアブストラクトは重要です。学会のプログラムにアブストラクトが含まれており，アブストラクトを見て学会の発表の場に参加者はやって来ますので，アブストの内容が明確で魅力的となるよう努力すべきです。学会発表用のアブストラクトは論文のアブストラクトよりも長いのがふつうです。発表用のアブストラクトはいくつかのパラグラフと1つあるいは2,3の図からなっています。

Chapter 11

Paraphrasing

Paraphrasing is a **restatement**, in your own words, of someone else's ideas. Paraphrasing is a very important skill.

paraphrase 言い換える
restatement 言い換え

11.1 Process of paraphrasing

To paraphrase a sentence, it is important to first recognize the main point of the sentence. Otherwise, it is very easy for the meaning of the sentence to be changed.

The following verbs can be used to state ideas:

*ask   maintain   assert   report   suggest   **claim**   deny*

claim 主張する

If you wish to search for good ways of paraphrasing, one of the most helpful methods is to compare two news articles reporting on the same event in two well-known newspapers. The stories will contain almost **identical** information; however, the sentences will be quite **distinctive**. By studying these sentences and paragraphs you will get better ideas about how to paraphrase.

identical 同じ
; ⇒ⓖ15.2 (3)
distinctive 特徴的

11.2 Paraphrasing strategies

There are specific approaches to changing sentences into your own words. The three main ways are:

(1) using different words that imply the same thing (a **synonym**)
(2) changing the order of the words
(3) changing the form of the word (for example, *developing* to *development,* verb to noun)

synonym 同義語

11.2.1 Use a variety of words that have the same meaning

Attempts to paraphrase words mostly utilize **substitution** by synonyms. However, the synonym that should be chosen has to depend on the context.

Most people try to use this technique, but it is difficult to master. The reason for this is that, while English words have many synonyms, which you can find by using a **thesaurus** or dictionary, these words rarely have the same exact meaning. So, if you attempt to paraphrase by using a large number of synonyms, there is a **considerable** chance that the sentences produced will not be in natural sounding English. Because of such unnatural and **peculiar** grammatically correct English, the reader in many cases may not be able to understand what you are trying to describe. The best way to use synonyms is to follow the "100% rule": only use a synonym for a word if you are absolutely certain that the new word has the same meaning.

substitution 置き換え

thesaurus 類語辞典

considerable かなりの
peculiar 奇妙な

> **A synonym** is a word or phrase that has almost the same meaning. For instance, *smart, clever* and *intelligent* are all synonyms since they all have basically identical meanings.

11.2.2 Change the order of the words

Another way to show your understanding of the original text is to reorder the information in the sentences while **retaining** the original meaning. However, deciding <u>which phrases to</u>

retain 保持する
which phrases to move
⇒ⓖ12.6

move or to which position the phrases must be moved is not always easy. Furthermore, in order to retain the original meaning of the sentence, you may need to add some phrases or words while deleting others to ensure that the new sentence is grammatically correct. Here are some ideas for changing the word order without changing the meaning:

(1) Use logical connectors, such as *although, because of, as a consequence of*

>   Example:
>   · The gravitational force is still not fully understood. However, the basis of our current understanding of it lies in Newton's law of gravitation.
>   →Although the gravitational force is still not fully understood, the starting point of our current understanding of it lies in the law of gravitation of Isaac Newton.

current ⇒W 4

(2) If the original sentence has two or more **clauses**, the order of the clauses can be interchanged.

clause 節

>   Example:
>   · With a little assistance, most people can paraphrase quite successfully.
>   →Most people can paraphrase quite successfully with a little assistance.

(3) If the original sentence possesses an **adjective** and a noun, the adjective can be replaced by a **relative clause** that provides the same meaning.

adjective 形容詞
relative clause 関係詞節

>   Example:
>   · Writing research papers is a difficult task for some people.
>   →Writing research papers is a task which is difficult for some people.

A clause is comprised of several **consecutive** words in a sentence that include a subject and a verb. A relative clause contains either a **relative pronoun** such as *who, which,* or *that*, or a **relative adverb** such as *when* or *where* at the start of the clause.

consecutive 連続した
relative pronoun 関係代名詞
relative adverb 関係副詞

11.2.3 Use grammatical variation

(1) Change a few of the words and phrases within the original sentence into other **parts of speech**.

a part of speech 品詞

(2) Often as a result the word order also will have to be changed as well as the insertion of some different words.

>   Example:
>   · Physicists like to study **seemingly** unrelated phenomena.
>   →Physicists like to **indulge** in studying seemingly unrelated phenomena.

seemingly 見たところ ⇒G 5.4
indulge 熱中する
active voice 能動態
and vice versa そしてその逆も真である ⇒G 16.9.9

(3) If the original sentence is in the **active voice**, it can be changed to passive **and vice versa**.

>   Example:
>   · You need to understand the forces acting on the **particle**.
>   →The forces acting on the particle need to be understood.

particle 粒子

Exercises

Q Choose the correct synonym for the word in *italic*:

i. The magnitude of the car's **constant acceleration** before *impact* was quite large.
   (a) force
   (b) shock
   (c) effect

constant acceleration 等加速度

Chapter 11

ii. Our *concern* is with situations in which the **constants** are zero.

    (a) company

    (b) anxiety

    (c) interest

iii. A quantum computer has the *potential* to simulate things that a classical computer cannot.

    (a) possible

    (b) capability

    (c) amount

constant 定数，常数

第11章　パラフレーズ

　パラフレーズとは，自分のことばで他の人のアイデアを述べることで，ぜひ身につけなければならないスキルです。

11.1　パラフレーズを書くプロセス

　ある一文をパラフレーズするには，まず，その文の要点を正確につかむことが重要です。それができないと，文意を曲げてしまうことになってしまいます。

　意見を述べるのに次のような動詞がよく用いられます。

ask　maintain（主張する）　assert（主張する）　report　suggest　claim（主張する）　deny

　パラフレーズの技術を身につけるには，1つの出来事を報じた2つの有力紙の記事を比較してみることです。情報はほぼ同じですが，文章は異なっています。個々の文と文章全体を対照することでパラフレーズのコツが得られます。

11.2　パラフレーズを書く方法

　文章を自分の言葉に変える具体的な方法があります。中でも重要なものに次の3つがあります。

(1)　同じ意味をもつ別の語（同義語）の使用

(2)　文中の語順の入れ替え

(3)　語の形の変換（例：developing → development，動詞→名詞）

11.2.1　同じ意味をもつ別の語の使用

　パラフレーズには同義語による置換がよく利用されます。しかしどの類義語を選ぶべきかは文脈によります。

　ほとんどの人がこのテクニックを使おうとしますが，使いこなすのは簡単ではありません。英単語には類義語が多くあり類義語辞典や辞書を使えば見つけられますが，これらの単語がまったく同じ意味をもつことはほとんどないからです。そのため，同義語を多用して言い換えようとすると，自然な英語とは言いがたい文章になる可能性が高くなってしまいます。そのような文法的には正しいけれども不自然で変わった英語表現のために，読者は著者が何を表現しようとしているのかを多くの場合理解できないかもしれません。元の語と置き換えられる語の意味が100%同一であると自信がもてるときのみ，置き換えるのです。

> 同義語とは，ほとんど同じ意味をもつ語（句）をいいます。smart, clever, intelligent は基本的な語義を共有しており，同義語といえます。

11.2.2　語順の転換

　原文が理解できていることを示すもう一つの方法は，元の意味を保ちながら文中の情報を並べ替えることです。しかし，「どの語句をどこに移すか」は，必ずしも容易な作業ではありません。加えて，元の文の意味を変えず，文法の規則を守った文を作るため，語句を加えたり，原文の語句を削ったりすることがあります。以下，具体例で説明します。

(1)　although や because of, as a consequence of などの論理コネクタ（結合語）を使う

    例：

    ・The gravitational force is still not fully understood. However, the basis of our current understanding of it lies in Newton's law of gravitation. (重力にはまだ解明されていないところがあるが，重力に対する私たちの現在の認識はニュートンの重力の法則に沿っている。)

→**Although** the gravitational force is still not fully understood, the starting point of our current understanding of it lies in the law of gravitation of Isaac Newton.

(2) 文が複数の節からなっていれば，前後を入れ替えることができる

例：

・With a little assistance, most people can paraphrase quite successfully. (少しヒントが与えられれば，大方の人はパラフレーズの技術を身につけることができる。)

→Most people can paraphrase quite successfully with a little assistance.

> 文の一部であり，主語と動詞をもつものを節と呼びます。関係詞節は，who, which, that のような関係代名詞や，when, where のような関係副詞を節の最初にもっています。

(3) 形容詞と名詞のある文では，形容詞を関係詞節に転換できる

例：

・Writing research papers is a difficult task for some people. (論文を書くのは必ずしも容易なことではない。)

→Writing research papers is a task **which** is difficult for some people.

11.2.3 品詞や態の変換

(1) 元の文中のいくつかの語（句）を他の品詞に変換する

(2) (1)に伴い，語順の転換や新しい語の付加が必要となることがある

例：

・Physicists like to **study** seemingly unrelated phenomena. (物理学者は一件無関係な現象に目を向ける。)

→Physicists like to indulge in **studying** seemingly unrelated phenomena.

(3) 元の文が能動態であれば受動態に変換でき，その逆も同様である

例：

・You need to **understand** the forces acting on the particle. (粒子に加わる力の性質を理解する必要がある。)

→The forces acting on the particle need to be **understood**.

Chapter 12

Concise technical writing

Most scientific and technical communications are generally relatively **concise**. It is because, if the author is a scientist, he or she will be writing largely for other scientists. Consequently, it is **critical** for a scientist to be **succinct**, **precise**, and **focused**.

Scientists rarely deal with isolated facts or events, therefore simple phrases are uncommon. The scientist must demonstrate relationships, including not just what occurs, but also how, when, and why it occurs, as well as what is affected. Also, the writing should be impersonal. The following factors <u>contribute to</u> a lack of **clarity** and conciseness:

concise 簡潔な

critical 極めて重要な
succinct 簡潔な
precise 正確な
focused 焦点をしぼった

contribute to ⇒ 🄶16.9.2
clarity 明快さ

- Long sentences
- Sentences that **reflect over-enthusiasm**
- **Verbosity** or **wordiness**
- **Redundancy** or the usage of terms with poor information content

reflect over-enthusiasm 過度に感情のこもった
verbosity 冗長な
wordiness 冗長な
redundancy 繰り返し，冗長性，重複表現

12.1 Reasons for conciseness

When writing, the fewest number of words possible should be used - this does not mean that the writing will contain less scientific substance, but rather that it is necessary to discover the clearest and most concise way to present the content. The reasons for conciseness may be as follows:

(1) Many publications, especially those with a large readership, set strict limits on the amount of words each article can have. Furthermore, certain papers, such as abstracts and **grant** proposals, have **stringent** limits on the number of words allowed. **CVs**, posters, and slides all have a **finite** amount of space.

grant 助成金
stringent 厳しい
CV 履歴書
finite 限られた
ambiguity 曖昧さ
identical 同一の

(2) Fewer mistakes will happen in the English text if the writing is concise, without redundancy or **ambiguity**.

(3) Avoiding **identical** words and phrases will also improve the readability of the article greatly.

12.2 Ways of writing concisely

(1) Avoiding nominalization is a good way to make writing more concise. Nominalization is the process of converting a verb or adjective into a noun.

Examples:
- The experience of worry for students with respect to being at university for the first time is common.
→Many students experience worries when they first go to university.

- The **enhancement** of flavor by the addition of a little orange **zest** is wonderful.
→A little orange zest enhances the flavor wonderfully.

enhancement 高めること
zest 風味

(2) Try to avoid adjective clauses which start with relative pronoun:

Example:
- The **light detector**, which measures frequency, is used in this experiment.

light detector 光検出器

Concise version:

・The light-frequency detector is used in this experiment.

(3) Use the most concrete word available wherever possible, avoiding **generic** words. Words like *activity* and *task* do not contribute anything; they are too **abstract**. The following are some more abstract words which you should attempt to minimize the use of:

character, condition, nature, situation, circumstance, instance, **eventuality**, **phenomenon**, problem, purpose, process, procedure, operation, phase, step, task, tendency, **intervention**, criteria, **facilities**, factor, realization, remark, activity, case

(4) Using verbs: Using more verbs reduces the quantity of words needed, enhances readability. Using a single verb (e.g. *analyze*) rather than a verb followed by a noun (e.g. *conduct an analysis*) is preferable.

avoiding ⇒Ⓖ11.2
generic 一般的な
abstract（形容詞）抽象的な

eventuality 不測の事態
phenomenon 現象（複数形
は phenomena）
intervention 干渉
facilities 設備
e.g. ⇒Ⓖ16.9.6

12.3 Redundancy

If a word or phrase adds nothing to the reader's understanding, then it is redundant. Readers will believe that if the initial few sentences of an article or paper suffer from redundancy, then the rest of the text will likely contain repetitions as well. As a result, instead of reading every single word, they will start scanning, or reading **one out of every five or six words**.

Everything written should be **of benefit**. To remove redundancy from your writing, it is not enough to simply delete some words. Consider removing entire sentences, paragraphs, or even subsections of text. The main points of the paper will **stand out** clearly for the reader if all **superfluous** things are **eliminated** and **accuracy is embraced** throughout.

When a **header** precedes an introduction sentence, you should carefully check it for redundancy. The following sentence, for example, if it appears directly after a heading titled Results, will be totally unnecessary: *The results of this work may be summarized as follows.* It is also unnecessary to begin a section titled Conclusions by writing: *In conclusion, we can say that ...*

When the word count of a sentence is reduced, the subject of the sentence is typically placed closer to the start of the sentence. This means that the reader receives **a much clearer picture** of the sentence content in a much less amount of time. Also, the reader will see more clearly the worth of what you are writing about if you use the fewest words possible.

one out of every five or six
words 5, 6 語のうち 1 語
of benefit 役に立つ

stand out 目立つ
superfluous [suːˈpɜːrfluəs]
余計な
eliminate 削除する
accuracy is embraced 正確
さを大切にする
header タイトル（章や節
の）
precede … ～に先行する
⇒Ⓦ15
a much clearer picture はる
かに明確な記述（picture
は絵ではなく，記述）

Table 12.1 Concise versions of long phrases

Phrase	Concise version
At all times	always
By means of	by
Due to the fact that	because
For the purpose of	for
In order to	to
In the nature of	like
In accordance with	according to

Exercises

Q1 Change the following verbs to nouns:

cite, communicate, conserve, develop, educate, inform, introduce, predict, produce, solve

Writing

Chapter 12

> **Q2** Write a paragraph on **radioactive** testing. Make it as concise as possible with the methods described in the text.

　radioactive 放射能

第12章　簡潔な英文作成

　一般的に，ほとんどの科学技術コミュニケーションは，比較的簡潔です。それは，著者が科学者である場合，他の科学者に向けて書くことが多いからです。したがって，科学者にとって，簡潔で正確かつ焦点を絞った文章を書くことは非常に重要です。

　科学者が，事実や出来事を他と切り離された単独のものとして取り扱うことは，めったにありません。したがって，簡単な表現ですませるということはほとんどありません。科学者は事実や出来事の関係を示さなければなりません。何が起きたかだけでなく，どのような現象がどのような環境下で，いつ，なぜ起きたのか，そしてどのような影響が生じたかを示します。さらに，論文は個人の観点から記すことはできません。

　文章の明確さや簡潔さを失わせる要素として次のようなものがあります。

- ・長い文章
- ・感情的な文
- ・冗長な（だらだらと綴られている）文
- ・重複表現，また，意味や情報をほとんどもたない語の使用

12.1　簡潔に書く理由

　論文は簡潔を旨としますが，といっても，科学的な内容に乏しい文章になるということではありません。そうではなく，内容を最も明瞭にかつ簡潔に書く方法を身につける必要があります。なぜ簡潔でなければならないか。次のような理由があります。

(1) 多くの出版物，特に広く読まれているものは各記事の語数に厳しい制限を設けている。また，アブストラクトや助成金の申請といった特定の書類は語数が厳密に定められている。履歴書やポスター，スライドなどはすべて，使えるスペースが限られている。

(2) 冗長さや曖昧さのない簡潔な英文を書けば，英語で書いたテキストにミスは少なくなる。

(3) 同じ概念や意味を表す同義表現を避けることで，文章は格段に正確に読みやすくなる。

12.2　簡潔に書く方法

(1) 簡潔な英文を書くのに，**名詞化を避ける**ことが役立つ：名詞化とは動詞や形容詞の名詞への転換です。

　例：

・The experience of worry for students with respect to being at university for the first time is common.

→Many students experience worries when they first go to university. （多くの学生が大学入学時は不安を覚える。）

・The enhancement of flavor by the addition of a little orange zest is wonderful.

→A little orange zest enhances the flavor wonderfully. （少しオレンジの皮を加えると風味は素晴らしく高まる。）

(2) **関係代名詞で始まる形容詞節の使用を極力を避ける**：

　例：

・The light detector, **which** measures frequency, is used in this experiment. （周波数を計る光検出器をこの実験で使う。）

　簡潔な表現：

→The light-frequency detector is used in this experiment.

(3) **一般的な語を避けてできるだけ具体的なことばを使う**：activity や task のような語は抽象的過ぎます。以下のような語はできれば避けましょう。

character（性格），condition（状態），nature（性質），situation（状況），circumstance（状況），instance（例），eventuality, phenomenon, problem（問題），purpose（目的），process（プロセス），procedure（手順），operation（操作），phase（段階），step（ステップ），task（タスク），tendency（傾向），intervention, criteria（基準），facilities（設備），factor（要因），realization（実現），remark（発言），activity（活動），case（ケース）

⑷ **動詞を用いる**：動詞を多く用いることで全体の語数を減らし，読みやすくなります。「動詞＋名詞」（例：conduct an analysis）ではなく動詞（例：analyze）を使うようにしましょう。

12.3 重複の表現

　新たな情報をもたらさない表現（語，語句）は冗長です。記事や論文の最初の 2, 3 文に重複表現が見られると，読者は残りの文章も同じように重複があるだろうと考えるでしょう。その結果，すべての単語を読むのではなく，スキャニング，すなわち 5, 6 語のうち 1 語しか読んでもらえないかもしれません。

　一語一句意味のある（何らかの情報を伝える）表現でなければなりません。重複の表現をなくすためには，一部の語を削るというだけではいけません。文章全体やパラグラフ，あるいはテキストの一部分を削除することを考えてください。余分なことは省き正確さに徹すれば，論文の要点が読者に明確に伝わります。

　第 1 文の前に見出しがついていたら，その第 1 文が重複していないか確認してください。たとえば，結果セクションの第 1 文に "The results of this work may be summarized as follows." と記す必要はありません。同様に結論のセクションを "In conclusion, we can say that …" で始めるのは無意味です。

　文の字数を減らすと，一般的に文の主語は文頭に近く配置されます。ということは，読者はより少ない時間で文の内容をより明確に受け取ることができます。また，できるだけ少ない語数で書けば，書かれていることの価値を読者はより明確に理解することができるのです。

表 12.1 簡潔な表現例

語句	簡潔な表現
at all times	always
by means of	by
due to the fact that	because
for the purpose of	for
in order to	to
in the nature of	like
in accordance with	according to

Chapter 13

Presenting information orally

People are expected to prepare content and convey it orally at university and in the workplace. Enhancing **oral** communication skills can be achieved through activities such as reading, refining presentation abilities, and regular practice.

oral 口頭の

13.1 Main points of a presentation

(1) **Preparation** - It is **critical** to **conduct** extensive **research** on the subject using a variety of sources. As soon as the topic has been decided, research and planning should start.

critical 極めて重要な
conduct research 研究を行う

(2) **Speaking style** - To keep your audience interested, you should speak slowly and clearly. The presentation should be practiced several times in front of a mirror.

(3) **Supplemental computer programs** - To deliver the presentation, nowadays computer applications like PowerPoint or Keynote are usually employed. The writing on the slides should be in a large font, so that they can be read easily from the back of a room.

13.2 Requirements of a presentation

The presentation must provide an opinion, which must be backed up by evidence from reading, research, and logical reasoning. From the presentation, the audience should be able to understand that the speaker is familiar with all the relevant literature. In addition, you must be able to demonstrate that you have critically analyzed the research you have conducted.

13.3 Visual aids of a presentation

Visual aids can help you improve your presentations by helping the audience grasp your topic, clarifying points, making an impact, and generating enthusiasm about your research. There are some tips to keep in mind when preparing a presentation file.

visual aid 視覚資料

- Because it is impossible for an audience to read and listen at the same time, too much writing on the slides will **detract from** what the speaker is saying. As a result, it is **vital** that the writing on the slides be brief, precise, and focused. It is not required to use whole sentences when writing. Only write down the most important points. It is acceptable to write in the style of a note with **bullet points**. **Indenting** is also a good tool for indicating which ideas are the most important.

detract from 損なう ⇒W12
vital 重要だ

bullet point 箇条書き
indent インデント

- "A picture is worth a thousand words" when it is both clear and accompanied by an appropriate **caption**. Depending on the topic, utilize photos, flowcharts, organization charts, tree diagrams, tables, or other visual representations.

caption キャプション

- It is crucial to ensure that the audience can see the slides clearly. Leave lots of 'white space' on your slide—that is, do not fill it entirely with text.
- The slides need to be structured in such a way that the audience's attention is focused on the point being made and there is an **anticipation** for what is to come next.

anticipation 期待
for ⇒G7.5

- Some references should be included in oral presentation slides, preferably one or two at a maximum on the particular **relevant** slide.

preferably ⇒G5.4
relevant 適切な

13.4 Body language during a presentation

Make motions and use gestures. Make eye contact with everyone in the room, but do not **stare** at them for too long. Make an extra effort to make eye contact with people on the far side of the room as well as those in the middle, as persons on the far side of the room are more likely to be **overlooked**.

stare じっと見る

overlook 見落とす

13.5 Overcoming nervousness

The greatest way to deal with nervousness before a presentation is to ensure that you are well prepared. Rehearse to ensure that the timing of the presentation is accurate <u>and that</u> there is also enough time to make <u>any necessary</u> **modifications**. During a presentation, most people speak too quickly because of natural **anxiety**. It is therefore necessary to practice speaking at a slightly slower pace than feels natural, but not too slowly.

<u>and that</u> ⇒G 6.1.2
<u>any necessary modifications</u> ⇒G 3.6
modification 修正（modify の名詞形）
anxiety 不安

Make a draft of the speech
Writing out the speech in English allows you to **decide upon** the proper grammar, vocabulary, style and tenses to use at different points in the presentation. At least 20% of the words and phrases used by untrained speakers are redundant, meaning they provide no meaningful information to the audience. Having a written speech during preparation of the presentation allows you to correct and **modify** as much as needed.

decide upon 決める（decide ではない。前置詞をとる）

modify 修正する

13.6 Inviting questions

The purpose of attending conferences is not only to present, but also to network with other researchers and academicians. The <u>question-and-answer session</u> is a good place <u>from where</u> the conversation can start. Furthermore, the questioners may be the same individuals who assist you in clarifying critical aspects of your research, or may want to collaborate with you or invite you to their laboratories.

<u>question-and-answer session</u> ⇒G 16.1
<u>from where</u> ⇒G 8.1

Answering complicated questions
Being well prepared is the greatest method to deal with **complicated inquiries**. Make a list of at least three <u>questions you expect</u> to be asked at the conclusion of your presentation. However, if a question is asked and you do not know the answer, there are many ways that you can politely and smartly admit this:

complicated 複雑な
inquiry 問いかけ（調査，探究の意味もある）
<u>questions you expect</u> ⇒G 8.3

- acknowledge that you do not know or are not sure;
- offer an idea of where and how the solution might be found;
- explain why you do not know, <u>e.g.,</u> no one knows;
- guess the answer but state clearly that it is just an **educated guess on your part**.

<u>e.g.</u> ⇒G 16.9.6
educated guess 知識や経験に基づく信頼度の高い推測
on your part あなたの側の
<u>et al.</u> ⇒G 16.9.8

Examples of answering complicated questions:
- That's a fascinating question, but I'm afraid I don't have an answer; however, I believe the answer may be found in the works of Tanaka <u>et al.</u>, published last year.
- That would be a fascinating subject for research! That hasn't been done yet, as far as I know.
- I'm not sure, but I'll do my best to find out.
- I believe it is around 100 in number, but I am not certain. **I'll look into it** for you.

I'll look into it 調べてみるね

13.7 Discussions

The importance of discussion in academic life cannot be **overstated**. You need to be able to start, add to, and participate in conversations about the issues raised by your **peers** from di-

overstate 誇張する
peer 同級生，同僚

Chapter 13

verse subjects.

Begin speaking

The only way to break through any mental obstacles is to speak up, even if it is sooner than you believe you are ready. Recognizing when another speaker is **pausing** or **hesitating**, as well as when it is appropriate for you to begin speaking, can be beneficial. You can also start speaking by simply agreeing or disagreeing with a speaker, indicating that **it is now your turn**.

pause（発言を）やめる
hesitate ためらう
it is now your turn あなたの話す番

Exercises

Presentation

Q1 Prepare a short presentation of around five minutes <u>on</u> an environmental problem. Prepare computer slides to accompany the presentation. The presentation should then be given to the class. It is very useful to comment on each other's presentations.

on ⇒ⓖ7.6

Q2 Prepare three questions and their answers on your presentation.

第13章　プレゼンテーション（口答発表）

　みなさんはこれから，大学や職場で研究成果を口頭で発表することになります。読書をし，プレゼンテーション能力を磨き，定期的な練習をすることにより，口頭でのコミュニケーション能力を高めることができます。

13.1　プレゼンテーションの要点

⑴　**準備**：さまざまな情報源を使い，テーマについて幅広くリサーチすることが重要です。トピックが決まったら早速，リサーチとプランニングに取りかかってください。

⑵　**話し方**：人を引きつける発表をするには，ゆっくり，はっきりと話すことです。鏡の前で数回発表練習を行うことを勧めます。

⑶　**補助的なコンピュータプログラムの活用**：プレゼンテーションにパワーポイントやキーノートなどが用いられます。部屋の後ろからでもよく見えるようスライドのフォントは大きくしてください。

13.2　プレゼンテーションに求められるもの

　プレゼンテーションでは，読書と調査，論理的推論による証拠に基づいた意見を述べなければなりません。発表を聴いている他の研究者が，「発表者は関連文献を熟知しており，今回の研究をすべて批判的に分析，検討したうえで発表している」と納得できる発表でなければなりません。

13.3　プレゼンテーションでの視覚資料の活用

　視覚資料の使用により，聴き手はトピックの把握が容易になります。また視覚資料はポイントを明確にし，聴き手にインパクトを与え，発表されている研究への強い関心を引き出します。以下，視覚資料を使う際のポイントを紹介します。

- ・読みながら同時に聞くことは聴き手にはできませんので，スライドに多くの文章を書いてしまうと，話し手の話の邪魔をしてしまいます。そのため，スライドに書く文章は簡潔に，正確に，焦点を絞って書くことが重要です。すべての文を書く必要はありません。最も重要なポイントだけを書いてください。箇条書きのメモのようなスタイルでも構いません。また，インデント（字下げ）も有効です。
- ・写真に曖昧なところが一切なく，適切な見出しがついていれば，"一枚の写真は1000語の力をもちます"。研究トピックによっては，写真やフローチャート，組織図，樹形図，表，その他の視覚資料も工夫してください。
- ・スライドがはっきりと見えるようにしてください。スライドを情報で埋め尽くすのではなく，余白を十分にとってください。

・他の研究者が発表に無理なくついていきこの次に何が提示されるかを予測できるよう，スライドを発表の流れ（論理構成）に沿って並べてください。
・関連するスライドごとに参照文献を，できれば2個以内で載せてください。

13.4 発表中の身振り手振り

発表では，身振り手振りを有効に使ってください。会場の全員に対して目を向けてください。ただし，一度に向ける時間は短めにしてください。真ん中で聴いている人だけでなく，後ろで聴いている人にも目を向けるよう気配りが必要です。遠くの人には目がいかなくなりがちです。

13.5 平常心を保つ

落ち着いて発表するためには，できることはすべてした，と思えるよう，入念な準備を行うことが必要です。制限時間内に発表を終え，かつ，必要であれば修正を加える時間も確保できるよう，発表練習を十分に行ってください。緊張のため発表は早口になりがちですので，ふつうのスピードよりいくらか遅く話す練習が欠かせません。ただし，遅すぎるのも問題です。

発表原稿の作成

スピーチを英語で書き出すことで，プレゼンテーションのさまざまな場面で使う文法や語彙，スタイル，時制を正確に用いることができます。訓練を受けていない話者が使用する単語やフレーズの少なくとも20%は冗長なものであり，それらは聴き手に意味のない情報を与えてしまいます。発表原稿を書くことで，必要に応じて修正することができます。

13.6 積極的に質問を受ける

学会に出席する目的は，学会で発表するだけでなく，他の研究者や学者と知り合うことです。知り合うきっかけは，発表後の質疑応答の時間にあります。質問者はみなさんの研究の核になるところを教えてくれるかもしれません。あるいは，みなさんに共同研究の提案をしたり，研究室に招いてくれたりすることがあるかもしれません。

難しい質問への答え方

入念な準備が，難しい質問への精一杯の対応です。発表の準備をする段階で，想定される質問を少なくとも3点挙げ，備えてください。

しかし，質問されて答えを知らない場合，礼儀正しくスマートにそれを認める方法はたくさんあります。

・知らない，またはわからないと認める
・答えがどこでどのように見つかりそうか，アイデアを提示する
・たとえば「誰も知らない」などわからない理由を説明する
・答えを推測するが，それは自分の推測に過ぎないことを明確に述べる

難しい質問への回答例：
・That's a fascinating question, but I'm afraid I don't have an answer; however I believe the answer may be found in the works of Tanaka et al., published last year. （すばらしい質問をありがとうございます。私に解答の用意はありませんが，昨年発表された田中氏らの論文に示されているかもしれません。）
・That would be a fascinating subject for research! That hasn't been done yet, as far as I know. （すばらしい研究トピックですね。私の知る限り，まだ手がつけられていないのではないでしょうか。）
・I'm not sure, but I'll do my best to find out. （よくわかりませんが，これから全力で取り組みます。）
・I believe it is around 100 in number, but I am not certain. I'll look into it for you. （100前後だと思いますが，確実とはいえません。これから調べます。）

13.7 議論

研究者・技術者として生きていくには議論の大切さは強調されなければなりません。さまざまなトピックについて研究仲間が提起した問題について会話を始め，補足し，参加することが求められます。

発言すること

不安を断ち切るただ一つの方法は，たとえ自分が準備できていると考えるより早くても思い切って発言することです。他の研究者が口をつぐんだり迷っているようなときに，またあなたが発言しても構わないと判断できたとき，思い切って発言してください。発表者の考えに賛成または反対の意見を述べるだけでもよいのです。そうすることで，発表の機会が回ってくるかもしれません。

Chapter 14

Marking stages of a presentation

Structure gives a presentation a framework. For many people, giving a presentation is a difficult **endeavor** that causes them a tremendous amount of anxiety. You will appear much more confident and **calmer** if you spend some time learning how good presentations are constructed and then applying that structure to your own presentation.

endeavor 努力
calmer 落ちついて

14.1　Main parts of a presentation

(1) Introduction - You should begin by introducing yourself and stating the **aim** of your presentation. The title of the lecture, background information, definitions of **terms**, and the **scope** of what you will cover are all included in the introduction.

aim 目的
term 用語
scope 範囲

(2) Outline - After greeting and introducing yourself to the audience, you must **clarify the substance** of your presentation. You can do this in several ways, but the most usual is to give a brief outline of the structure of the presentation. Use connectors or signaling language to accomplish this. A presentation, <u>unlike writing</u>, lacks paragraphing to indicate when the speaker is shifting topics or completing his or her remarks. As a result, this signaling language is important throughout the major body of the presentation and will aid the audience in understanding and appreciating your work.

clarify the substance 内容を明確に述べる

unlike writing ⇒Ⓖ16.3

(3) Body - The body of your paper <u>is where</u> you display your research, make **pertinent** points if you have an argument to make, **contradict** other arguments, and list your references.

is where ⇒Ⓖ8.2
pertinent 適切な
contradict 反論する ⇒Ⓦ6

(4) Conclusion -<u>This is where</u> the main points are summarized, and future recommendations are made.

This is where ⇒Ⓖ8.2

Connectors: Connectors are **conjunctive words** in English grammar that are used to connect similar elements in a sentence. Intelligent usage of connectors eliminates boring single sentences and lends <u>coherence</u> to what you write or say orally. Aside from joining words, sentences, or paragraphs, sequence connectors can help you organize your thoughts when you want to tell someone what happened at a **specific** time **in a logical sequence**.

conjunctive word 接続語
coherence 首尾一貫性
⇒Ⓦ17

specific 特定の
in a logical sequence 論理的な流れで

Signaling language: In an oral presentation, (1) you must explain the topic to your audience, (2) identify the main sections of your speech, and (3) connect your ideas and information so that (4) the presentation flows. This can be accomplished using signaling or transition words. Signaling words are words that allow a listener to predict what will happen next and what relationships exist between ideas. Signaling words in conversations thus function similarly to roadside signposts.

14.2　Phrases used in different parts of a presentation

Discourse markers are words and phrases that we employ to organize and connect our thoughts. They function as signposts, indicating to the listener what information will be presented next. The audience is thus guided through the presentation via signpost language. This is critical since a presentation without a defined structure will lose the audience's attention. It is also crucial to use a signal when you are **wrapping up** your presentation to let people know **you are done**.

wrap up 締めくくる
you are done 発表が終了した

(1) Introduction

The title and introduction of the presentation are similar to an **advertisement**: you want as many people to be interested in your research as possible, so it should not be too technical or **generic**. At least 30% of the effectiveness of your presentation is determined by how you introduce yourself and how the audience reacts to your introduction. It takes around 90 seconds for an audience to establish an opinion about a presenter, after which it is difficult to change their minds.

advertisement 広告

generic 一般的な

State your plan

· What I'd like to do is talk about ...
· What I want to do is convey ...
· In today's talk, I ...
· Today's topic is ...
· Today, I'm going to discuss ...
· I'm going to speak with you about ...
· I will give a brief lecture <u>on</u> ...
· Today I want you to think about ...
· In this presentation, I want to focus <u>on</u> ...
· The topic of this presentation is ...
· The goal/aim/objective of this presentation is to ...
· This presentation is intended to ...

state 述べる

on ⇒ G 7.6

on ⇒ G 7.6

State how you will carry out the talk

· I'm going to cover three elements of the topic ...
· My presentation will be divided into ... sections.
· My presentation is organized into ... parts.
· I decided to break up my presentation into ... pieces.
· This topic can be researched under the following headings: ...
· I'm going to take approximately ... minutes.
· The presentation should last about ... minutes.
· At the end, I'll be pleased to answer any questions.
· If you have any queries, I'll do my best to respond later.
· Please feel free to interrupt if you have any questions.

(2) Main body

Determine how much material you can offer in the time **allotted** after you <u>have defined</u> the goal of your presentation. Also, take advantage of your knowledge of the audience to design a presentation that is detailed enough. You do not want to create a presentation that is either too simple or too complex.

allotted 与えられた
⇒ G 10.1
have defined ⇒ G 9.1.3

Ordering points

· First and foremost, ...
· To begin with, ...
· Secondly, ...
· Thirdly, ...
· Finally, ...

Using examples

· As an example, ...
· For example, ...
· As evidence of this, consider the following: ...
· Keep in mind ...
· All you have to do is consider ...

Emphasizing

· In addition, ...
· Furthermore, ...
· This backs up my **claim** that ...
· As a result, **it follows that** ...

claim 主張
it follows that ... ～という
ことになる

Chapter 14

In reference to what has been said
- As I stated at the **outset**, ...
- In the first half of my presentation, I stated ...
- As I already stated, ...
- I just informed you a few minutes ago ...

outset 初め

Putting it in another way
- To put it another way, ...
- To put it this way, ...
- The point I'm trying to make is ...
- What I'm proposing is ...

put it in another way
言い換える

Referring to visuals
- This graph shows ...
- Take a look at what's going on here.
- Let's take a closer look at this.
- Please have a look at this.
- I'd like to call your attention to the following ...
- As you can see, ...
- The ... **stands for** ...
- The graph **depicts** ...
- As may be seen from the graph, ...
- If you look **attentively**, you'll notice

stand for ... ～を表している
depict ... ～を表している
attentively 注意して

Moving on to the next point
- Now I'd like to move on to ...
- Now we'll have a look at ...
- Now we'll go on to ...
- After looking at ..., I would like to/want to think about ...
- Let's move on to ...
- Now I'd like to move on to ...
- The following point is ...
- Another important topic to consider is ...
- The second point I'd like to discuss is ...
- That brings me to ...

(3) Conclusion
A standard and effective way to conclude a presentation will include the words *thank you*, but it is also good if the concluding slide contains a graphic (ideally one that summarizes the entire presentation or refers to an earlier slide) as well as **contact information**.

contact information 連絡先

- So ...
- We've seen this before ...
- We began by looking at ... and discovered that ...
- Then we debated ... and I argued ...
- In summary ...
- **In a nutshell**, we've looked at ...
- To summarize ...
- Finally, I would like to emphasize that ...
- That, **I believe**, covers most of the points.
- My presentation is now complete.
- Thank you for your time and consideration.
- **That's all there is.** If you have any suggestions or questions, please let me know.
- That concludes my major points.
- Is there anything else you'd like to suggest or ask?
- I'd be happy to address any queries you may have.

in a nutshell 一言でいえば

I believe ⇒Ⓖ16.3

That's all there is. 発表はこ
こまでです。

> The term "audience" refers to a group of people. It is singular if you think and/or express it as a group; plural if you think and/or express it as individuals acting within the whole.

14.3 Phrases used during the question-and-answer session

Positive feedback

When giving a presentation, presenters, especially students, may feel uncomfortable and exposed. It is critical that the audience offer them positive feedback so that they can believe that they will be able to produce a reasonably competent presentation sooner or later.

Examples:
- I really enjoyed how you explained the ...
- I really liked how you made it clear ...
- Your slides helped me understand your presentation because ...
- The presentation was fascinating because ...

Exercises

Presentation

Q1 Complete the sentences used in different stages of a presentation with words below:

Aim, begin, elaborate, finally, problem

i. The _____ of this presentation is to tell you about a new airplane.

ii. I'd like to _____ by asking you to think about something.

iii. So, what is the _____ with traditional designs?

iv. And _____, I'd like to _____ on the advantages of our design.

Q2 Think about how you can improve the design of a technological object (airplane, robot, computer, for example). Prepare a presentation detailing your hypotheses and discuss the advantages and disadvantages of your idea.

第 14 章　効果的なプレゼンテーション

　構成はプレゼンテーションに枠組みを与えます。プレゼンテーションをするにはたいへんな努力が必要で，大きな不安にかられます。すぐれたプレゼンテーションの構成を学びそれを生かすことに時間をかければ，もっと自信があり落ち着いてみえるでしょう。

14.1　プレゼンテーションの主要部分

(1)　序論

　始めに自己紹介を行い，プレゼンテーションの目的を述べます。タイトル，背景の情報，用語の定義，取り上げる範囲はすべて序論に含まれます。

(2)　アウトライン

　自己紹介の後，プレゼンテーションの目的を述べることから始めるべきです。これにはいろいろな提示法がありますが，最も一般的なのは，プレゼンテーションの構成のアウトラインを示すことです。そのためには，コネクタやシグナリング語となる言葉を使いましょう。文章とは異なりプレゼンテーションでは，話し手がトピックを移したり発言を終えたりするタイミングを示すパラグラフがありません。そのため，プレゼンテーション全体を通してシグナリング語が重要であり，その言葉は聴き手が発表者の研究を理解し，評価する助けとなります。

(3)　本論

　研究の全容を示します。論じる内容を適切に述べ，他の研究の問題点を指摘します。また，参

考文献を提示します。

⑷　結論

研究の要点を述べ，聴き手への提案を行います。

> 💡 **コネクタ**：コネクタとは英文法における接続語のことで，文中の類似した要素をつなぎます。コネクタをうまく使うことで，退屈な単文をなくし，書いた文章や口頭での発言に一貫性をもたせることができます。単語や文，パラグラフをつなげるだけでなく，特定の時間に起こったことを論理的な順序で誰かに伝えたい場合にも，first, lastly などのシークエンス・コネクタは考えを整理するのに役立ちます。
>
> **シグナリング語**：口頭プレゼンテーションで求められるのは，⑴トピックを聴衆に説明する，⑵発表の主要な部分を明確にする，⑶自分の意見とそれを支える情報を結びつけること，それによって⑷発表をスムーズに進行させることです。それを可能にするのは思考の流れをスムーズに示すことば（シグナリング語や転換語）です。これを適切に用いることで，聴き手は次に何が起こるか，アイデア間にどのような関係が存在するかを予測できます。シグナリング語は，道端の道標と同じように機能するのです。

14.2　プレゼンテーションのさまざまな場面でよく用いられる表現

discourse marker とは，自分の考えを整理し，つなげるために用いる語句のことです。次に提示する情報を聴き手に示す道標として機能します。この discourse marker（signpost language）により，聴き手はプレゼンテーションの展開を容易に把握できます。discourse marker はとても重要です。それがなければプレゼンテーションは構造を欠いたものとなり，聴き手は聴こうという気持ちをそがれてしまいます。プレゼンテーションを締めくくる際にもそれを知らせる表現を使うことが大切です。

⑴　序論

タイトルとプレゼンテーションの序論の部分は広告に似ています。できるだけ多くの人に研究に関心を持ってほしいので，専門的すぎても一般的すぎてもいけません。プレゼンテーションの成否の 30% を発表者の自己紹介とそれに対する聴き手の関心が握っています。最初の 90 秒で，このプレゼンテーションは聴くに値するかどうかを聴き手は判断し，それを過ぎると聴き手の態度を変えさせるのは困難です。

切り出しのことば

- What I'd like to do is talk about …
- What I want to do is convey …
- In today's talk, I …
- Today's topic is …
- Today, I'm going to discuss …
- I'm going to speak with you about …
- I will give a brief lecture on …
- Today I want you to think about …
- In this presentation, I want to focus on …
- The topic of this presentation is …
- The goal/aim/objective of this presentation is to …
- This presentation is intended to …（本発表の目的は～）

どのように議論を進めるか述べる

- I'm going to cover three elements of the topic …
- My presentation will be divided into … sections.
- My presentation is organized into … parts.
- I decided to break up my presentation into … pieces.
- This topic can be researched under the following headings: …
- I'm going to take approximately … minutes.（およそ～分をかけます。）
- The presentation should last about … minutes.（プレゼンテーションは約～分で行います。）
- At the end, I'll be pleased to answer any questions.
- If you have any queries, I'll do my best to respond later.
- Please feel free to interrupt if you have any questions.（発表の際中に質問がありましたら遠慮なくどうぞ。）

（2）　論文の本体

　プレゼンテーションの目的を話した後，限られた時間で話す内容を決めておいてください。また，話す内容に聴き手がどの程度通じているのかがわかるのであれば，それに応じて話の難易度を調整してください。プレゼンテーションは聴き手に，易しすぎても難しすぎても困ります。

論の展開
- First and foremost, ...
- To begin with, ...
- Secondly, ...
- Thirdly, ...
- Finally, ...

例の提示
- As an example, ...
- For example, ...
- As evidence of this, consider the following: ...
- Keep in mind ...（〜に留意してください。）
- All you have to do is consider ...（ただ一つ，〜についてお考え下さい。）

強調
- In addition, ...
- Furthermore, ...
- This backs up my claim that ...（これは，〜という私の主張の支えになります。）
- As a result, it follows that ...（その結果〜ということになる。）

すでに述べたことへの言及
- As I stated at the outset, ...（最初に述べましたように）
- In the first half of my presentation, I stated ...
- As I already stated, ...
- I just informed you a few minutes ago ...

言い換え
- To put it another way, ...（言い換えますと〜）
- To put it this way, ...
- The point I'm trying to make is ...
- What I'm proposing is ...

グラフや表などを示す
- This graph shows ...
- Take a look at what's going on here.
- Let's take a closer look at this.（もっと詳しく見てみます。）
- Please have a look at this.
- I'd like to call your attention to the following ...
- As you can see, ...
- The ... stands for ...（〜を表しています。）
- The graph depicts ...（このグラフは〜を描いています。）
- As may be seen from the graph, ...（グラフからおわかりいただけるかもしれませんが，〜）
- If you look attentively, you'll notice ...（注意して見ていただければ，お気づきになるでしょう。）

次のトピックに移る
- Now I'd like to move on to ...
- Now we'll have a look at ...
- Now we'll go on to ...
- After looking at ..., I would like to/want to think about ...
- Let's move on to ...
- Now I'd like to move on to ...
- The following point is ...
- Another important topic to consider is ...
- The second point I'd like to discuss is ...
- That brings me to ...

（3）　結論

　プレゼンテーションを終えるスタンダードかつ効果的な方法は感謝のことばを伝えることです

が，最後のスライドに連絡先の情報だけでなくグラフィック（理想的にはプレゼンテーション全体を要約するか，それまでに使用したスライドを示すもの）を含めるのもよいでしょう。

- So ...
- We've seen this before ...
- We began by looking at ... and discovered that ...
- Then we debated ... and I argued ...
- In summary ...
- In a nutshell, we've looked at ...
- To summarize ...
- Finally, I would like to emphasize that ...
- That, I believe, covers most of the points. （本発表の重要な部分にほとんど触れたと思います。）
- My presentation is now complete.
- Thank you for your time and consideration. （ご清聴ありがとうございました。）
- That's all there is. （発表はここまでです。）If you have any suggestions or questions, please let me know.
- That concludes my major points. （本発表の主要な部分でここまでです。）
- Is there anything else you'd like to suggest or ask?
- I'd be happy to address any queries you may have. （ご不明な点がございましたら，気軽にご質問ください。）

> "audience"という人の集まりを表す語は，1つのグループとしては単数形で，グループの個々の成員に着目する時には複数形で扱われます。

14.3　質疑応答の時間に用いられる表現

敬意を示す

　発表者（特に学生の場合）は極度に緊張し，人前に無防備にさらされたような気分になりがちです。聴き手は発表者に肯定的なフィードバックを与え，発表者が遅かれ早かれそれなりのプレゼンテーションができるようになるだろうと自分で信じられるようにすることが求められます。

　例：

- I really enjoyed how you explained the ...
- I really liked how you made it clear ...
- Your slides helped me understand your presentation because ...
- The presentation was fascinating （すばらしい）because ...

Chapter 15

Explaining technology to non-specialists

It is vital to be able to convey technical knowledge via understandable language. You may gain trust with your audience and develop a shared understanding if you do it properly.
If you are required to **impart** technical knowledge to a non-technical audience:

(1) Do not make any **assumptions**.
(2) Do not damage your efforts from the start by thinking that your target audience is <u>tech-savvy</u>.
(3) Before the meeting, do a quick survey of the audience to **identify** their ability level.
(4) Do not employ **jargon** from the technical world. Except when absolutely required, avoid **acronyms**.
(5) When you <u>do</u> need to use technical vocabulary, make sure you first describe it in plain English.
(6) Before moving on, double-check that everyone understands the words.

> While both *adopt* and *accept* mean to receive as one's own an opinion, policy, or practice, *adopt* implies accepting something created by another or unfamiliar to one's nature.

Make sure you only give enough information to highlight the most important points when explaining something technical. You do not want to **delve** so deeply **into** the specifics that you **overwhelm** or lose your audience.
Make sure to slow down when introducing unfamiliar technical ideas to your audience. Scientists tend to forget the audience might be deeply confused about the topic. So, instead of rushing through your topics, you should talk more slowly than is natural. Allow yourself more time than you think you will need.

> **Tips to keep in mind:**
>
> · Be aware of your audience.
> · Recognize their concerns.
> · Respond to their issues.
> · Avoid using jargon.
> · Make use of simple **metaphors**.
> · Make a drawing.
> · Do not get too involved with the details.
> · Speak slowly.
> · Use two or more explanations.
> · Consider your audience's viewpoint.
> · Pay attention to the audience's body language.

15.1　Phrases to signal explanation

Transition words and phrases can assist the reader or listener to understand the logic of your research by creating effective ties between concepts. Some of the words or phrases used to signal intention of clarifying things are:

· another way of putting this is that
· i.e.

impart 伝える

assumption 仮説

tech-savvy 技術に精通している ⇒ G 16.1
identify 確認する
jargon 専門用語（専門外の人にはわからない，と批判的に使われることが多い）
acronym 頭字語（ET など）
do ⇒ G 16.6

delve into … 〜を掘り下げる
overwhelm 圧倒する

metaphor 比喩

- in other words
- or
- that is
- that is to say
- to put that in **layman's** terms
- to put this in everyday terms

layman 素人

Examples:
- This is another way to define the **linear momentum** of a system of particles.
- It can generate 550 nm (**i.e.**, visible) electromagnetic waves.
- In other words, the results are extraordinary.
- The Earth's climate is intricately connected to various factors, such as greenhouse gas concentrations and solar radiation, that is, the complex interplay of these elements shapes our planet's weather patterns.
- That is to say, the left half of the system accounts for 10% of the initial intensity.
- This heat must be added to **vaporize** a liquid or must be removed to condense a gas.

linear momentum 線形運動量

account for ⇒ G 16.9.1
vaporize 気化させる

15.2 Metaphors

A popular way for explaining things to non-specialists is the use of **metaphors**. They are an excellent example of using a **circumstance** that everyone is familiar with to convey an abstract technical idea in sufficient depth so that the audience can grasp all of the main points without being lost in technical jargon. Metaphors are a fantastic tool to have in your technical communication toolkit.

metaphor 比喩
circumstance 状況，場面

Using **analogies** related to your audience's field is a wonderful way to construct a metaphor. Analogies help us understand unfamiliar concepts by likening them to something more familiar. Wherever possible, technical jargon should be avoided and when it is unavoidable it needs to be explained simply. Remember experts may not find simplistic examples instructive, and that non-specialist readers may completely **miss** highly technical examples. It is essential to ensure that your audience is familiar with the subject of your metaphor regardless of their background, experience, culture, where they grew up, or whatever topic in which they are engaged. To explain technical terms or ideas to a non-technical or non-specialist audience it often helps to make metaphors with the following:

analogy 類推

miss 見逃す，気がつかない

- Parts of functions of the human body
- Everyday objects
- Everyday actions and events

Example of metaphor:
- The concept of electricity can be likened to the flow of water in a river.

Example of analogy:
- Just as water flows from higher to lower elevations, electricity moves from areas of high electrical potential to areas of lower potential.

This analogy helps visualize the behavior of electric currents and understand phenomena like resistance and voltage.

Avoid sports metaphors. Not everyone is familiar enough with sports to fully comprehend the metaphor or to see how it might be used to illustrate a message. Furthermore, sports language is competitive and individualistic, which are not traits you want to promote when helping people learn something new.

15.3 Other ways to explain things simply

Rather than repeating the same thing again and again, employ a variety of ways to teach your subject. Instead of using too many words, make a graphic instead. You can **invigorate** your audience by just displaying crucial and relevant information in a format they can readily understand and digest by producing visualizations with your data. Only use numbers if they are directly related to anything the audience need to know.

Finally, when you are speaking, make sure you are paying attention to your audience and their body language. It may be necessary to finish the presentation a little bit early when confronted by a lot of **yawning**, sleepy eyes, or **slouching**, instead of **persisting** when the audience is tired.

invigorate 活気づける

yawning あくび
slouching しゃがむ姿勢
persist … あくまで〜しよう
とする，〜し続ける
⇒W11

Exercises

Writing

Q1 Use the following words and phrases to write complete sentences showing an explanation:

Another way of putting this is that; *i.e.*; *in other words*; *or*; *that is*; *that is to say, to put that in layman's terms*; *to put this in everyday terms*

Q2 Choose any **principle** or technique that is relevant to your field of research. Explain it in two separate paragraphs, i. one technical and ii. the other non-technical.

principle 原理

第 15 章　一般の人々への技術の説明

専門知識をわかりやすいことばで伝えることは重要です。適切に行えば，聴き手の信頼が得られ，聴き手との間に共通の理解を得ることができるようになります。

専門家以外の聴き手に発表する際に重要なことは，以下です。

(1) 仮説を立てない
(2) 聴き手は技術的なことに精通していると考えて，発表に至るまでの努力が最初から台無しになるようなことはしてはいけない
(3) 発表の前に，聴き手の理解のレベルを知るために簡単なアンケートを行う
(4) テクノロジー業界の専門用語を使わない。どうしても必要な場合以外は，頭字語は避ける
(5) 専門用語を使用する際には，その用語の平易な説明をしておく
(6) 次に進む前に，全員が専門用語を理解できたか確認する

adopt も accept も，提案された意見や方針，慣習を受け入れることを意味しますが，adopt は他の人が考えた，あるいはなじみのない意見や方針，慣習を受け入れるということを意味しています。

専門的な内容を説明するときには，最も重要な点にしぼって話をしてください。細部に入り込み聴き手の関心を失うことは避けなければなりません。

聴き手になじみがないような内容を説明するときには，話す速度をゆるめてください。話しているトピックについて聴き手がかなり混乱しているかもしれないことを科学者は忘れがちです。トピックを急いで話すのではなく，普段話すときの自然な流れよりもゆっくり話すべきです。必要だと思う以上の時間をとってください。

発表に際しての留意点：
・聴き手を意識する
・聴き手の関心や不安を理解する
・聴き手の問題意識（関心事項）に答える
・特定の分野の専門用語を使わない
・わかりやすい比喩を使う

Chapter 15

```
・図を用いる
・細かい議論に入り込まない
・ゆっくり話す
・説明を 2, 3 取り入れる
・聴き手の視点（考え）を考慮に入れる
・聴き手の体の反応を見逃さない
```

15.1　説明を示す表現

　概念間の効果的な結び付きをつくることば（シグナリング語句や転換語句）を使うことで，読み手や聴き手は研究の論理を理解しやすくなります。物事を明確にする意図を示すのに用いられる語句に次のようなものがあります。

- another way of putting[*1] this is that（別な言い方をすれば）
- i.e.（すなわち）[*2]
- in other words（別のことばを使うと）
- or（あるいは，こうもいえる）
- that is（すなわち）[*3]
- that is to say（それは，以下のようにいうこと（と同じ）だ，すなわち）
- to put that in layman's terms（一般の人でもわかることばでいえば）
- to put this in everyday terms（日常のことばを使っていえば）

[*1] put＝express
[*2] i. e.＝that is
[*3] that is＝that is to say

例：

- This is **another way** to define the linear momentum of a system of particles.（粒子システムの線形運動をこのように定義することもできる。）
- It can generate 550 nm (**i.e.,** visible) electromagnetic waves.（これにより 550 nm の（すなわち可視レベルの）電磁波が生まれる。）
- **In other words,** the results are extraordinary.（言い換えると，結果は並外れたものだ。）
- The Earth's climate is intricately connected to various factors, such as greenhouse gas concentrations and solar radiation, **that is**, the complex interplay of these elements shapes our planet's weather patterns.（地球の気候は，温室効果ガス濃度や太陽放射などさまざまな要素と複雑に関係して生じている。つまり，これらの要素が複雑に絡み合って地球の気象パターンを形作っている。）
- **That is to say**, the left half of the system accounts for 10% of the initial intensity.（すなわち，システムの左半分は初期強度の 10% を占めている。）
- This heat must be added to vaporize a liquid **or** must be removed to condense a gas.（この熱を加えて液体を気化させる，あるいは，この熱を除いて気体を濃縮する必要がある。）

15.2　比喩

　一般の人々に物事を説明するのによく使われる方法として比喩の使用があります。比喩は誰もがよく知っている状況を利用して，抽象的な技術的アイデアを十分に伝え，聴き手が専門用語に困惑せずに要点をすべて把握できるようにした優れた例です。比喩は理工系の分野のコミュニケーションにもツールキットとして大いに役立ちます。

　比喩で説明する際，聴き手にわかりやすいアナロジーを用いると効果的です。アナロジーは，なじみのない概念を別のなじみのあるものになぞらえることでその概念の理解を容易にしてくれます。専門用語は極力避けるべきで，やむを得ず使う場合でも，平易な言葉で説明してください。専門家にはわかりやすい例は必要ないでしょうが，一般の人には，高度に技術的な例を示されてもついていけないことを忘れずにいてください。みなさんの用いる比喩表現を聴き手が理解できるよう工夫してくだい。彼らのさまざまな背景や経験，文化，育った環境，関心のある領域の違いにかかわらず，専門家の用いる比喩表現を理解できるようにしてください。専門用語やアイデアを一般の人に説明するのに，以下のような比喩が役に立ちます。

- 人体の各部位と機能
- 身のまわりのもの
- 普段の行動や出来事

比喩の例：

- Electricity is like the flow of water in a river.（電気は川の水の流れに似ている。）

アナロジーの例：

- Just as water flows from higher to lower elevations, electricity moves from areas of high electrical potential to areas of lower potential.（水が高い所から低い所へ流れるように，電気は高い電位から低い電位へと動く。）

　このアナロジーにより電流の動きを視覚的にとらえることができ，抵抗や電圧といった現象が
理解しやすくなります。

　スポーツに関する比喩は避けてください。多種のスポーツがあり，どのスポーツの比喩を使う
にせよ，そのスポーツになじみのない人が多くいるからです。それに，スポーツの語彙は競争や
個人に関係しており，新しい研究内容を説明するのには向きません。

15.3　その他の簡単な説明法

　同じことを繰り返していうのではなく，さまざまな表現を試みてください。ことばだけでの説
明でなく視覚情報も活用します。重要で関連性の高い情報を，聴き手がすぐに理解して消化でき
るように視覚情報を用いて表示すれば，聴き手を引きつけることができます。数字のデータは，
聴き手が知る必要がある場合にのみ提示しましょう。

　最後に，話すときには聴き手の顔の表情や体の動きに目を止めてください。聴き手がしきりに
あくびをするとか，眠そうな目になる，体が傾くことがあれば，疲れている聴き手を相手に発表
を続けるのではなく，はやめに切り上げる必要があるかもしれません。

Part 2

Effective strategies for improved writing and presentation using academic science texts

Preface – Part 2

Writing and presentation are important tools for <u>effectively</u> describing <u>one's</u> research and also conveying to the audience an understanding of the learner's skills, passions and knowledge. Knowing your audience is the first and most **crucial** rule of presenting your work. You will have a great presentation if you can comprehend your listeners' needs and interests. Therefore, throughout the preparation of a presentation, **be it** oral **or** written, the different aspects of language that make a subject matter interesting to an audience should be kept in mind.

In Part 2, students approach English study through interesting texts that are connected to science and technology. This gives them the opportunity of applying what they learn to their work and academics immediately. Effective strategies for improved writing and presentation are also systematically explained in the chapters of Part 2, accompanied by explanations of particular linguistic skills. There are exercises for the students to properly understand the material. The vocabulary and important expressions taught in the chapter is thus revised. As a result, students should be able to better present their own ideas concerning a related topic by engaging their critical thinking.

Almost all of the text has been translated into Japanese. Reading a book in two languages at the same time is a wonderful method to develop language skills and gain confidence when students realize how much of the language they already know. The different phrases of the English language become easier to comprehend when compared with the styles and phrases of your **native tongue**.

We believe that this book will be able to equip learners with the necessary English linguistic skills necessary to function successfully in their future work environments in our modern interconnected world.

effectively ⇒ G 5.2
one's ⇒ G 3.2
crucial 非常に重要な

be it A or B A であれ B であれ（ = whether it is A or B）

native tongue 母語

第 2 部
効果的な論文作成と発表―理工系諸分野のテキストを読みながら―

第 2 部への序

　論文を書き，そして発表することは，みなさんの研究成果を効果的に表現し，みなさんがもっている技術や研究への情熱，知識を受け手に伝える重要な手立てです。研究の成果を発表する際にまずもって重要なのは，受け手について理解していることです。聴き手が必要としていることや関心をもっていることを理解できていれば，発表の質が高まります。そのため口頭であれ論文であれ，プレゼンテーションの準備全体を通して，論じるトピックが受け手を引きつけるものとなるよう，論文の執筆と発表の際に求められる言語のさまざまな側面を心に留めておく必要があります。

　第 2 部では，みなさんが興味をもてる理工系のテキストを通して技術英語を学びます。第 2 部の学習を通して，学んだことをすぐに仕事や研究に応用できる機会が出てきます。第 2 部の各章では，効果的に論文を書き口頭で発表するための方策が個々の技術の解説とともに順序立てて説明されています。練習問題でその章の内容をきちんと理解できたかを確認し，語彙や重要表現を復習できます。結果としてみなさんは，批判的思考を働かせることで研究しているトピックについての成果をより効果的に発表できるようになるはずです。

　テキストはほぼすべて日本語に訳されています。一つのトピックを英語と母語で学ぶことにより，英語のテキストに対するみなさんの理解の程度がわかり，英語での論文発表に対する自信が得られるようになります。母語の文体や表現と比較することで，英語のさまざまな表現が理解しやすくなります。

　本書はみなさんが将来国境を越えた場で活躍する際に必要となる，英語のスキルを提供すると，私たちは思っています。

Chapter 16
Explaining necessity

Students need to develop the ability to give clear instructions and explain the purpose and outcomes of various projects. By using concise and **lucid** language to confirm the necessity for a particular research goal, the researcher will be more likely to produce a high-quality technical document and thus promote more <u>positively</u> a future research course.

lucid わかりやすい

<u>positively</u> ⇒Ⓖ5.4

16.1 Language for explaining necessity

Have to, *need to*, *must*, *should* and *ought to* are the **modal verbs** (words which express the speaker's attitude) used to talk/write about a necessity in the present or future. *Should* and *must* appear most frequently in scientific writing, usually in the passive voice. *Must* expresses a very strong need, <u>anything else being</u> impossible. *Have got to* is used in informal English.

modal verb 助動詞

<u>anything else being</u> ⇒anything の前に with を補うと，付帯状況を表す構文Ⓖ7.2 として理解できる

Example:
・There **must** be some record.

Should is a milder term compared to *must* and can at times convey a suggestion or recommendation.

Example:
・The establishment of this simple rule **should** be enough.

However, in technical English, *should* and *must* can sometimes be interchanged.

16.1.1 Difference in necessity

(1) Somebody else other than the speaker has made the decision

Examples:
・He **has to** check the result of the experiment.
・This is a danger which **needs to** be **averted** by **transpiration**.
・This **ought to** solve his problems.

avert 回避する
transpiration 蒸散

(2) The speaker decides that something is necessary

Example:
・I **must** take all responsibility.

16.1.2 Difference in negativity

(1) *Must not* does not mean the same as *do not have to*. If you *must not* do something, that means it is important that you not do it. On the other hand, if you do <u>not</u> have to do <u>something</u>, it means that it is not necessary for you to do it, but you can do it if you want.

<u>not ... something</u> ⇒Ⓖ3.5

(2) You use *need not have* and a past participle to express that something wasn't necessary, but still done.

Example:
・You **need not have** phoned them; they had been informed beforehand.

; ⇒Ⓖ16.2.1

(3) The modal verb *must* does not have a past form, so to express past obligation or neces-

sity, we use the modal verb *had to*. When you need to say that it was important that something did not happen or was not done, then other expressions have to be used, e.g., "ít was important not to". Phrases such as "had to make certain" used in a negative sentence can also provide the same meaning.

Example:
・We **had to** make certain that it was not **obsolete**.

obsolete 廃止となった

Text: Science and Scientists

The word "science" **is derived from** the Latin "scientia" meaning knowledge. Science is a way of learning about what exists in the natural world, how it functions, and how **it came to be the way it is**. In science, an investigative <u>question</u> must <u>be posed</u>, then original data **needs to** be generated and collected and finally a conclusion that answers the question **should be** made based on **available evidence**.

As astronomer Carl Sagan wrote: "Science invites us to let the facts in, even when they don't **conform to** our **preconceptions**. It counsels us to carry alternative hypotheses in our heads to see which <u>ones</u> best **match the facts**."

A <u>fact</u> is something that is true or something that has occurred or has been proven correct. A <u>concept</u> is defined as a general idea of something. A <u>principle</u> is a basic truth or the source or origin of something or someone. A <u>theory</u> is an idea to explain something, or a set of guiding principles.

Students of science can choose to **major in** various fields, such as mathematics, biology, physics, **calculus**, chemistry, geology, astronomy, statistics, or engineering. Some institutions use new research in their teaching, and include students in their **R&D**, thus enhancing both their teaching and education. **Faculty** and staff at universities and institutes **need to** assist students in achieving their academic and professional goals. Research advisors or supervisors are scientists who **have to** guide their **protégés** in the development of effective research methods and quantitative analyses.

is derived from ... 〜に由来する
it came to be the way it is 今の形になった
question ... be posed ⇒ⓖ16.5
available evidence 明確な証拠
conform to ... 〜に従う
preconception 先入観
one ⇒ⓖ3.2
match the fact 事実と適合する

major in 専攻する
calculus 微積分
R&D 研究開発
faculty 教授陣

protégés [próʊʈəʒèı] 守るべき人。ここでは研究指導を受けている人（学生）

Exercises

Grammar

Q1 Add *must* or *should* and use the passive form of the verb in parentheses.

i. A PCR test _____ (only administer) when necessary.

ii. Soap _____ (prepare) by heating glycerol and oleic acid together.

iii. The power to the **furnace** _____ (always turn off) before maintenance.

furnace 溶鉱炉

iv. The new safety protocol dictates that theseal on the **apparatus** _____ _____ (never release) by untrained personnel.

apparatus [ˌæpəˈrætəs] 装置

v. **Potassium** _____ (never allow) to come in contact with water.

potassium カリウム

Technical vocabulary

Q2 Put the following words in the spaces provided.

Science, Faculty, Research, Available, Evidence

i. There are many **renowned** scholars in the departments of _____ and technology.

ii. He graduated from the university's Chemistry _____ .

iii. My supervisor informed me when he would be _____ for weekly meetings.

iv. My current _____ topic requires running lab experiments on mice.

v. _____ of my research progress was provided to my professor.

renowned 著名な

Writing

Q3 Write a paragraph discussing a computer operating system. <u>In particular</u>, note the problems or features that need to be improved for such a structure. Present one or two ideas that can solve such problems. Use at least four modal verbs or modal phrases.

in particular ⇒ 🅖5.5

第 16 章　必要性の説明

　学生が身につけなければならないのは，明確に指示を与え，携わっているさまざまなプロジェクトの目的と結果を説明できる能力です。簡潔で明快なことばで特定の研究目標の必要性を確認することにより，研究者は質の高い技術文書を作成でき，それにより，研究者・技術者としての将来の道を切り開くことができます。

16.1　必要性を説明する表現

　have to, need to, must, should, ought to は 助動詞と呼ばれ，現在または未来のある行為に対する筆者の意見（すべき，すべきでないなど）を示します。科学論文で最も多く用いられるのは should と must で，いずれも受動態を伴うことが一般的です。他の可能性を排除する must はとても強い意見を表します。have got to はインフォーマルな場面で用いられます。

　例：
　・There **must** be some record. (記録がなければならない。)

　must に比べて should は穏やかなことばで，提案や推奨を伝えることもできます。

　例：
　・The establishment of this simple rule **should** be enough. (この単純な規則を立てておけば十分なはずだ。)

　しかし技術英語では，should と must が意味の違いなく用いられることがあります。

16.1.1　必要性の違い

(1) 発言者以外の人が決定した場合

　例：
　・He **has to** check the result of the experiment. (実験の結果を彼は確認しなければならない。)
　・This is a danger which **needs to** be averted by transpiration. (この状態は危険であり，蒸散させて危険を除く必要がある。)
　・This **ought to** solve his problems. (こうすれば彼の抱えている問題は解決されるはずだ。)

(2) 何かが必要であると話し手は判断する場合

　例：
　・I **must** take all responsibility. (責任は私一人で負うしかない。)

16.1.2　否定の程度の違い

(1) must not と do not have to は意味が異なります。must not は「〜してはいけない」の意味をもつのに対し，do not have to は「〜しなくてもよい（したければしてもよいが）」の意味をもちます。

(2) need not have + 過去分詞は「〜する必要はなかったのに（した）」という意味を表すのに使います。

　例：
　・You **need not have** phoned them; they had been informed beforehand. (何も電話なんかしなく

たってよかったのに。（そのことを）とっくに知っていたんだから。）

(3) 助動詞 must には過去形がないため，過去の義務や必要性を表現するには，must ではなく had to という助動詞を使います。何も起こらなかった，または行われなかったことが重要であると示す場合は，"it was important not to" などの他の表現を使用する必要があります。また，否定文の中で使われる "had to make certain" のようなフレーズも同じ意味を表します。

例：

・We **had to** make certain that it was not obsolete.（そのことが廃れてはいないことを示さなければならなかった。）

テキスト：科学と科学者

　"科学" という語は知識を意味するラテン語の "scientia" に由来します。科学は，自然界に存在するものとその機能，そして，それが今のかたちになった経緯（歴史）を研究します。科学では，まず予備的な問いを立て，それに関する新しいオリジナルデータを作り収集する，そして，信頼できる証拠に基づいて問いに対する解答を提出しなければなりません。

　天文学者カール・セーガンは「科学は私たちの常識では受け入れがたいことも事実として受け入れるよう私たちを誘います。いくつかの仮説の中でどれが一番事実に合致するか，対立仮説を頭の中に思い浮かべることをすすめます」と言っています。

　事実（fact）とは，真実であること，あるいは起こったこと，あるいは正しいと立証されたことを指します。概念（concept）とは個々の事実を越え一般性をもったものであり，原理（principle）とは，基本的な真実，または何かまたは誰かの出所または起源です。理論（theory）とは，何かを説明する考え方，あるいは，主要な原理から構成されるものです。

　自然科学を研究する人間は，自然界のさまざまな事象を研究の対象とします。たとえば数学や生物学，物理学，微積分学，化学，地質学，天文学，統計学，工学などです。研究機関の中には，教育に新しい研究を取り入れ，学生を研究開発（R&D）に参加させることで教育の質を高めているところがあります。大学や研究機関の教授陣やスタッフは，学生が研究目標を達成できるよう支援しなければなりません。指導教員は，学生に有効な研究法や定量分析の技術を身につけさせなければなりません。

Chapter 17

Expressing needs, problems and solutions

Expressing needs, and also the problems known, provides the **rationale** for research and employs data and other facts to **substantiate** the need for finding a solution to the problem. Critical thinking must be used to find the best solutions and then it has to be explained why your solutions are the best. To identify the problem and solution in a text, the reader should be able to easily find signal words and phrases such as:

rationale [ˌræʃəˈnæl] 根拠
substantiate 実証する

The **problem** / **issue** / **challenge** is; To address this; To deal with this; The solution suggested.

problem 問題
issue 課題
challenge 困難な課題

17.1 Expressing needs

We can view the concept of expressing needs as being divided into the three following categories:

(1) need to do something
(2) **need to <u>not</u> do <u>something</u>**;
(3) no need

need to <u>not</u> do <u>something</u>
する必要のないものがある⇒🄶3.5

We can also **regard** the idea **from the point of** view of the person/situation causing the need, and the person **addressing** the need. Here are some examples of English expressions that show necessity.

regard A from the point of
B B の観点から A をみる
address 対応する

Table 17.1 Verbs that help to express needs

	Very strong	Strong
Need someone to do something	**compel**, demand, force, **oblige**, must	require, need, have to
Need someone not to do something	ban, forbid, prohibit, must not, not be allowed to, not be permitted	cannot, may not, need not

compel 強制する
oblige 強制する

To express needs, the following words and phrases are helpful:

should; ought to; should/ought to be able to; should/ought to be capable of; should/ought to have the ability to

---

Difference between ban and prohibit:

To ban is normally the action of forbidding something by some law or social rule.
To prohibit is the action of not allowing something; it does not necessarily imply any rule or law.

Expressing objective

The following way of expression is commonly employed to state the objective of a study or research.

The aim The objective The purpose	of	the study the investigation	is was	to	understand discover find out	if whether

Chapter 17

17.2   Identifying a problem

Research is frequently described as a <u>problem-solving</u> activity, and as a result, descriptions of problems and solutions are an important part of scientific communication used to explain research activities.

problem-solving ⇒ G 16.1

Table 17.2 Structures for formulating a problem

	Possibility	Certainty
Present	*may/might/could/should* + *be* + adjective/**present continuous**: Example: · This assumption **may be invalid**.	*must* + *be* + adjective/present continuous: Example: · The gravitational force **must be** working against **the driving force**.
Past	*may/might/could* + **present perfect**: Example: · A **meteorite may** have altered Earth's climate.	*must* + present perfect: Example: · It **must** have been destroyed.

present continuous 現在進行形

invalid 無効だ
driving force 駆動力
present perfect 現在完了

meteorite [ˈmiːtiərait] 隕石

17.3   Suggesting solutions

Ways of suggesting solutions to problems are to use the following words:

(1) *propose/suggest/recommend* + *that* – suggesting the people or objects which willl help to solve the problem

Example:
· It is proposed that a spaceship may be <u>propelled</u> in the solar system by radiation pressure.

propel ⇒ W 7

(2) *propose/suggest/recommend* + *-ing* – suggesting the action needed to solve the problem

Example:
· We recommend installing a more secure system.

Text: Ethics

Decisions about right and wrong **permeate** everyday life. **Ethics** is a set of behavioral standards that helps us decide how we **ought to** act. In some ways, we can say that ethics is all about making the right decisions.

Science and technology have had a major impact on society as science has allowed us to take control of our species' development. The public sees science and technology as **net positives** for society. However, they have not been without **controversy**. According to some, technology creates more problems than it solves. Science has also become identified with death and destruction. The negative effects of the digital age will almost certainly have additional negative consequences. Bill Gates has said **of** artificial intelligence or A.I. that it is like nuclear energy — 'both promising and dangerous'. The ethical consequences of such advances should thus be **thought through**.

Ethical principles provide guidance for researchers in their professional, scientific and educational roles. <u>Accordingly</u>, it is recommended that the ethics of science and technology education and development for the betterment of humanity be **thoroughly** researched. Achieving this goal will **take time**, but it will never happen unless we make a start. Recognizing <u>one's</u> social responsibilities as a scientist is an important first step.

permeate 浸透する
ethics 倫理学

net positive 光を多くもたらす
controversy 議論，論争

of ... 〜について

be thought through 熟考される
accordingly ⇒ G 5.3
thoroughly [ˈθɜːrəli] 徹底的に
take time 時間がかかる
one's ⇒ G 3.1

Researchers are ethical agents who are responsible for the consequences of their actions. Instead of **withholding** relevant information, ethics requires that **authentic** results be reported. That is, scientists are expected to be truthful. Scientific responsibility is thus the duty to **conduct** and apply science **with integrity**.

> withhold 控える
> authentic 信頼できる，本物である
> conduct 推進する
> with integrity 誠実に

To develop: to grow or to become more advanced
Research: careful and organized study or gathering of information about a specific topic

Exercises

Understanding the text
Q1 Choose the answer *a*, *b* or *c* which best fits the explanation given in the text:
The lack of ethics of scientists and engineers can result in.

(a) lack of knowledge in the sciences.

(b) **crucially** important new discoveries and projects.

> crucially 決定的に

(c) adverse outcomes arising from the **application** of scientific and technological advancements.

> application 応用

Technical vocabulary
Q2 Match the following words from the left column to the phrases in the right column.

(1) responsible	(a) for the consequences
(2) application	(b) of research
(3) science	(c) of innovations
(4) fields	(d) of exploration

Grammar point
Q3 Complete the sentences by choosing the correct word from the following:

supposed, prohibited, must, banned

i. The government has _____ the dumping of waste chemicals without a license.

ii. Containers with **flammable** materials _____ be at least 50 meters from the building.

> flammable 可燃性の

iii. Walls are _____ to be **free of dirt**.

> free of dirt 汚れがない

iv. Fire **extinguishers** are _____ from being used near or on electrical equipment.

> extinguisher 消火栓

Writing
Q4 Explain the factors that have to **be taken into consideration** when a nuclear power plant is to be built and the need for nuclear energy use. Based on this information, assess the **rationality** for building a new nuclear power plant and write a report explaining your recommended action.

> be taken into consideration 考慮される
> is to ⇒ 12.7
> rationality 合理性

第 17 章　ニーズ，問題および解決策の表現

　ニーズ（必要なこと）や直面する問題を述べることで，研究の根拠が明らかになり，データや事実を利用して問題の解決策を見つける必要性が示されます。批判的思考により最良の解決策を見つけ，その解決策が最良である根拠を説明する必要があります。テキストに記された問題と解決策を，次のようなシグナリング語句に注目することで，容易に見つけることができるはずです。

the problem / issue / challenge is; to address this; to deal with this; the solution suggested.

17.1　ニーズを表す英語表現

ニーズを表す英語表現は 3 種類あります。

(1)　need to do something（する必要がある）

(2)　need not do something

(3)　no need

　ニーズを表す英語表現を，ニーズをもたらす人間（状況）や説く人間の立場からみることもできます。ニーズを表す英語表現の例を挙げます。

表 17.1　ニーズを表現する動詞

	とても強い	強い
人が何かを行う必要がある	compel, demand, force, oblige, must	require, need, have to
人が何かを行ってはならない	ban, forbid, prohibit, must not, not be allowed to, not be permitted	cannot, may not, need not

　ニーズを表すのに，次の語や句が役立ちます。

　should; ought to; should/ought to be able to; should/ought to be capable of; should/ought to have the ability to

> 💡 **ban と prohibit の違い：**
> ban：法や規則による禁止
> prohibit：禁止ではあるが，必ずしも法や規則による規制は伴わない

目的を表す表現

　以下の表現を用いて研究の目的を述べます。

The aim The objective The purpose	of	the study the investigation	is was	to	understand discover find out	if whether

17.2　問題の特定

　研究は問題解決の活動とよくいわれます。そのため，問題と解決策の記述は，研究活動を説明するのに用いる科学的コミュニケーションの重要な要素です。

表 17.2　問題を定式化するための構造

	かもしれない／しれなかった	ちがいない／ちがいなかった
現在	may/might/could/should + be + 形容詞 / 現在進行形： 例： ・This assumption **may be** invalid.（この仮定は誤っているかもしれない。）	must + be + 形容詞 / 現在進行形： 例： ・The gravitational force **must be** working against the driving force.（重力が駆動力を妨げているのはまちがいない。）
過去	may/might/could + 現在完了： 例： ・A meteorite **may** have altered Earth's climate.（隕石が地球の気候変動に関わったかもしれない。）	must + 現在完了： 例： ・It **must** have been destroyed.（破壊されたにちがいない。）

17.3　解決策の提案

　課題に対する解決策の表現に次のようなものがあります。

(1)　propose/suggest/recommend ＋ that – 問題の解決に役立つ人や物の提案

　　例：
　　　・It is **proposed that** a spaceship be propelled in the solar system by radiation pressure.（放射圧により太陽系で宇宙船を飛ばすという案が出されている。）

(2)　propose/suggest/recommend ＋ –ing – 問題の解決に必要な行動の提案

　　例：
　　　・We **recommend installing** a more secure system.（より安全なシステムの導入を勧める。）

テキスト：倫理

　「これは正しいのかどうか」の決断に迫られることは毎日のように起きています。その際に倫理学は，人間の行動の規範（基準）としてはたらきます。倫理学の本質は正しい行動の決定，決断にあります。

　科学技術は社会に大きな影響を与えてきました。科学によって人類はヒトという種の発展をコントロールできるようになったからです。人々は科学と技術が社会に役立つと信じています。しかし，これらの科学技術がいつもそのまま受け入れられてきたわけではありません。科学技術のもたらした益より害が多いとみなす人もいます。科学を死や破壊につながるものとする見方もあります。今日のデジタルの時代では，さらに科学の負の面が出てきています。ビル・ゲイツは人工知能（AI）を核エネルギーに類するものとみています。「明るい未来を切り開くと同時に，危険をはらんだ未来をもたらしかねない」といっています。そのような進歩のもたらす負の結果は何かを深く考えねばなりません。

　倫理原則は研究者に専門的，科学的，教育的役割において行動規範を示します。その規定に沿って，よりよき社会に向けた科学者・技術者の倫理教育が真剣に模索されなければなりません。効果ある倫理教育は容易には実現できませんが，始めないことには何も生まれません。まず，科学者の社会的責任を自覚することです。研究者は自分の行動の結果に対して責任のある倫理的主体です。関連情報を伏せるのではなく，倫理的な観点から本当の確かな結果を報告することが求められます。すなわち，科学者は真実であることが求められています。したがって，科学者の責任，義務とは科学を誠実に実施し適用することです。

to develop（**発達する**）：成長する，またはより高度になる
research（**研究**）：特定のトピックに関して慎重かつ組織的に研究し情報を集めること

Chapter 18

Hypothesizing

A **hypothesis** (**plural** hypotheses) is an explanation proposed for a phenomenon. The scientific method requires that a hypothesis be a specific, testable prediction of a study's outcome. Scientists generally **base** scientific hypotheses **on** previous observations that cannot be explained satisfactorily by the available scientific theories. It is therefore stated at the beginning of the study. <u>Typically</u>, this **entails** proposing a possible relationship between two variables: the independent variable (what the researcher modifies) and the dependent variable (what the researcher measures).

hypothesis 仮説
plural 複数形
base A on B B に基づいて A を打ち立てる
typically 通常 ⇒ G5.4
entails 伴う

18.1 Hypothesizing

A research undertaking usually attempts to solve a unique question, termed a research question. To do this, the researcher **first speculates as to** the answer to a problem: this statement is called a hypothesis. After this, he/she tries to find out whether his/her guess is right or wrong by either performing a theoretical calculation or experimental measurements.
Science starts with problems. Conclusions are **inferred** from a **tentative** hypothesis, which is proposed and then tested to be able to provide <u>an explanation for</u> the facts. A hypothesis does not have to be correct, and it may need to be rejected altogether or at the least revised.
The first step in the scientific method is thus to form a hypothesis. Scientists employ the scientific method to make observations, form hypotheses, and gather evidence in a theoretical or practical experiment designed to support or **refute** the hypothesis. As evidence is accumulated to support a hypothesis and it becomes popular within the scientific community, it is referred to as a theory.

first speculate あらかじめ見通しを立てる
as to ... 〜に関して

infer 推論する
tentative 仮の
<u>an explanation for</u> ⇒ G7.5

refute 否定する

First, firstly and at first: We frequently use *first*, particularly in writing, to indicate the order of the points we want to make. When making a list, we can use either *first* or *firstly*. *Firstly* is more formal than *first*. *At first* means "at the start" or "in the beginning," and it is used to contrast two things.

Examples:
· First(ly), the sodium chloride is dissolved in the water.
· At first, many students are homesick when they go abroad, but at the end of their studies they find themselves not looking forward to going back.

18.2 Expressing a hypothesis

A hypothesis is based on the existing scientific knowledge, **intuition**, and experience of the researcher. Even though it is possible to hypothesize about something that happened in the past or will happen in the future, most hypotheses are stated in the present simple tense. A hypothesis is typically written as an if/then statement. It is sometimes expressed as a prediction, using the future tense with *will*. To express a hypothesis or **speculation**, one of the verb forms listed in Table 18.1 is used.

intuition 直感

speculation 推測

Table 18.1

Tense	Verb form
Past	*Would/could/may/might/should* + past participle

Present	*Would/could/may/might/should* + main verb
Future	*Would/could/may/might/should* + main verb

Examples:
- By now, the train should have left. (past)
- She could arrive soon. (present)
- By next Monday, I may have created the required graphic. (future)

18.2.1 Verb forms in conditionals

A hypothesis can also be stated by using an *if* clause, sometimes called a **conditional clause**.
If clauses use different tenses from the main clauses. Table 18.2 shows the structures when dealing with a hypothetical situation in the present, past and future.

conditional clause 条件節

Table 18.2

Situation	'If' clause	Main clause	Examples
Past	Past perfect	*Would/could/may/might/should* + *have* + past participle	If they had replaced the turbine bolts at the time of installation, accidents might have been prevented.
Present	Past	*Would/could/may/might/should* + base verb	If they replaced the turbine bolts at the time of installation, accidents would not happen.
Future	Past	*Would/could/may/might/should* + base verb	If they replaced the turbine bolts at the time of installation, it should prevent accidents.

Usually, *was* is not used with conditionals. Instead, *were* is used for all cases.

18.2.2 Expressing probability

A hypothesis is usually **tentative**; it is an assumption or suggestion. There are many ways to express probability, from very strongly (e.g. *must*) to gradually indicating a very slight chance (e.g. *unlikely*).
Modal verbs can be used to suggest possibility, such as *must, may, might,* and *could*. Scientists often use **modals** to imply uncertainty, especially **in conjunction with** the word *be*.

tentative 仮の，一時的な

modal 助動詞
in conjunction with ... 〜と一緒に用いて

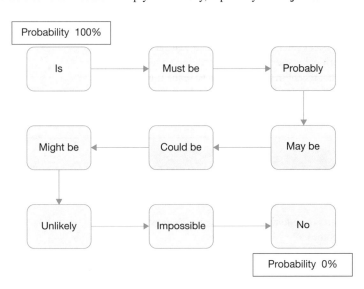

18.3　Testing a hypothesis

To test the hypothesis, before **embarking on** real research **comprised of** theoretical or experimental calculation, it is best to prepare about 5 to 8 questions whose answers will help to find out the answer to the main research question.

Of course, you may find that it is not possible to draw conclusions from the answers to several of the questions. You may also realize that these questions may not be appropriate for finding out all of the information required to solve your research question.

embark on ... 〜に実際に取りかかる，〜に乗り出す
comprised of ... 〜から成る

Hypothetical: a probable effect of an action, which has not happened yet

Text: Horizons in physical sciences

The physical sciences include four extensive areas: astronomy, physics, chemistry, and the Earth sciences. They deal with the systematic observation of **inorganic compounds**, **as distinct from** the observation of the natural world, which is the domain of biological science.

Physical science is now challenged by the <u>need for</u> its use in technological applications. Research is thus concentrated in mainly two areas: studying the properties of already discovered materials in order to obtain additional information which may lead <u>eventually</u> to an improved or cheaper product, and the production of absolutely new materials. Until a century ago, physical scientists celebrated the Newtonian model, <u>emphasizing</u> certainty over uncertainty and seeking elegant mathematical theories and neat solutions. However, **over the last century**, scientists have realized that the world is not like that — that the real world includes chaos, instability and uncertainty — and have thus developed **a plethora of** techniques for dealing with these realities, the most **prominent** <u>of which</u> is quantum mechanics.

In theoretical science, efforts are being made to **integrate** quantum mechanics with Albert Einstein's general theory of relativity, which serves as a framework for gravity. The search for the **elusive** unified theory has **given rise to novel** and **intriguing** concepts such as superstring theory. Computational science now allows us to investigate simulations of pattern formation in traffic flows and **pedestrian** motion, the behavior and interactions of different combinations of substances, and even bacterial cellular development, with the possibility of discovering new and innovative applications.

Whether the subject is a simple fluid or climate models with components such as the atmosphere, oceans, or greenhouse gases, physical scientists have learned that interactions between the different components are critical and cannot be directly linked to the properties of the individual things. As a result, a recent trend is to emphasize the importance of **interdisciplinary** physical science in order to benefit from the best ideas in various sciences.

inorganic compound　無機化合物
as distinct from ... 〜とは異なり
need for ⇒ G 7.5

eventually 最終的に ⇒ G 5.4

emphasizing ⇒ G 11.2

over the last century　過去100年を通して
a plethora of 多くの
prominent 目立った
of which ⇒ G 8.1
integrate 統合する

elusive とらえどころのない
give rise to 生み出す
novel 新しい
intriguing 興味深い
pedestrian 歩行者

interdisciplinary
[ˌɪntərˈdɪsəplɪneri] 学際的な

Exercises

Understanding the text

Q1 Choose the answer *a* or *b* in the following pairs of sentences to indicate the sentence that can be most considered as expressing a hypothesis:

i. (a) Hammers and golf balls fall faster than feathers.

  (b) An object made of wood may fall at a different **velocity** than one made of iron.

velocity 速度

ii. (a) The big bang theory was introduced in 1927 by Georges Lemaitre.

(b) The cosmic egg is the tiny, dense mass from which the universe may have exploded into existence.

Q2 From the text on *Horizons of Physical Science*, choose two hypotheses of scientists in the past century.

Grammar point

Q3 Underline the words that indicate probability in these sentences:

i. Scientists have discovered a strange phenomenon known as time reversal, in which light waves seem to travel backward in time.

ii. The **outer crust** is believed to be comprised of **nuclei** that are **densely** packed.

iii. The universe appears to be expanding faster than it should.

iv. Physicists say it might be possible to devise a theory of everything.

> outer crust 外部地殻
> nuclei 核（単数形は nucleus)
> densely 密に

Q4 Change the following sentences into hypotheses, using the words in brackets. Do not use the words in italics.

i. The rail *likely* suffered thermal shock because of excessive heat. (may)

ii. The aircraft *possibly* did not have a mechanical failure. (might not)

iii. The houses *surely* did not **collapse** after the earthquake due to design **blunders** (cannot).

> collapse 倒壊する
> blunder 誤り

Writing

Q5 Select a subject in which you are interested. Then, **decide on** a research question. Afterwards, write a statement which presents a hypothesis of the possible answer to this question. Prepare a **questionnaire** with about 5 and 8 questions to investigate all aspects of your research question and thus test your hypothesis.

> decide on ... ～を決定する (decide + 名詞とはならない)
> questionnaire [ˌkwestʃəˈner] アンケート

Chapter 18

第 18 章　仮説を立てる

　仮説はある現象に対する説明です。研究の方法が科学的であるためには，仮説は，研究結果を具体的かつ検証可能な形で予測するものでなければなりません。一般に科学者は既存の理論では十分に説明できない現象に基いて，仮説を立てます。したがって，仮説は研究の始めに述べられます。仮説の設定は通常，独立変数（研究者が修正するもの）と従属変数（研究者が計測するもの）の関係を提示する形で行われます。

18.1　仮説を立てる

　研究は通常，リサーチ・クエスチョンと呼ばれる独自の疑問を解決しようとする試みです。そのために研究者はまず，問題に対して仮説と呼ばれる解答を推測します。その後，理論上の計算や実験での測定を行い，その推測が正しいか誤っているかを判断しようとします。

　科学は問題から始まります。結論は暫定的な仮説から推論されます。提案されたその仮説は事実の説明ができるように検証されます。仮説は正しいものでなければならないということはありません。完全に否定されることもあれば，修正が必要という場合もあります。

　このように，科学的方法の最初のステップは仮説を立てることです。科学者は科学的な仮説手法を用いて観察を行い，仮説を立てます。そしてその仮説を支持または否定できるように設計された，理論上の，あるいは実際の実験により証拠を集めます。仮説を支持する証拠が蓄積され，それが科学界で一般的に受け入れられると，その仮説は理論と呼ばれるようになります。

> **first, firstly, at first:** first，firstly，at first：言いたいことの順番を示すために first を，特に文章を書くときによく使います。リストを作るときは，first と firstly のいずれも使います。firstly は first よりもフォーマルなことばです。at first は最初は，最初にという意味で，2 つの事柄を対比するときに使われます[*1]。
>
> *1 例：
> 最初は～だった。その後…となった。

例：
- **First** (**ly**), the sodium chloride is dissolved in the water.（最初に，塩化ナトリウムが水に溶ける。）
- **At first**, many students are homesick when they go abroad, but at the end of their studies they find themselves not looking forward to going back.（海外に行くと初めのうちはホームシックになる学生が多いが，帰る頃にはもっといたいと思うことが常だ。）

18.2　仮説を述べる

　仮説は研究者の科学的知見と直感，経験に基づきます。過去に起きたことやこれから起きるかもしれないことについての説であっても，大半の仮説は現在形で述べられます。仮説は，「もし〜であれば，…となる」の形で述べられます。仮説は，未来形（will）を用いて予測として述べられることもあります。仮説や推測を表すのに，表 18.1 に挙げた動詞の形のいずれかを用います。

表 18.1

時制	動詞の形
過去	would/could/may/might/should ＋ 過去分詞
現在	would/could/may/might/should ＋ 本動詞
未来	would/could/may/might/should ＋ 本動詞

例：
- By now, the train **should** have left.（過去）（今頃は，電車は出発しているはずだ。）
- She **could** arrive soon.（現在）（もう少しすれば着くはずだ。）
- By next Monday, I **may** have created the required graphic.（未来）（次の月曜日には，お求めのグラフィックができているかもしれません。）

18.2.1　条件法での動詞の形

　仮説は if 節（条件節）を用いて述べることもできます。if 節は主節とは異なる時制をとります。表 18.2 は仮定の状況を現在・過去・未来形で表現したときの文の構造を示しています。

表 18.2

状況	if 節	主節	例	
過去	過去完了	would/could/may/might/should ＋ have＋ 過去分詞	If they had replaced the turbine bolts at the time of installation, accidents **might have** been prevented.（タービンのボルトを取りつけ時に取り換えていたら，事故は防げたかもしれない。）	
現在	過去	would/could/may/might/should ＋ 原形動詞*2	If they replaced the turbine bolts at the time of installation, accidents **would** not happen.（タービンのボルトを取りつけ時に取り換えれば，事故は防げるだろうが。）	*2 あるいは動詞の不定形
未来	過去	would/could/may/might/should ＋ 原形動詞*3	If they replaced the turbine bolts at the time of installation, it **should** prevent accidents.（取りつけ時にタービンボルトを交換すれば，事故を未然に防ぐことができるはずだ。）	*3 あるいは動詞の不定形

　一般的に，条件節では was を were で置き換えます。

18.2.2　可能性の表現

　仮説は仮のものであり，仮定ないし示唆です。強い可能性を示す must から可能性がほとんどないことを表す unlikely まで，可能性の表現はさまざまです。

　must や may，might，could などの助動詞を用いて可能性を表すこともできます。科学者は助動詞を多用し，不確実性を表します。特に to be を伴います。

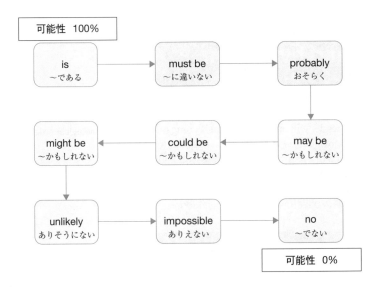

18.3　仮説を検証する

　仮説を検証するには，実際のあるいは理論上の実験での計算に入る前に，5 つないし 8 つの問いを立てるとよいでしょう。それに対する解答が研究での本格的な問いへの解答を見い出すヒントになることが期待されるからです。

　無論，それらの問いに解答を与えても仮説に対し何らかの結論は得られないかもしれません。また，これらの問いは本格的な問いに答えを出すのに必要な情報を得るのに適切ではないかもしれません。

hypothetical（仮説の上での）：これまで起きたことはない，ある行為から生じうる結果

テキスト：物理科学の地平

　物理現象を扱う科学は 4 つの広範な領域から構成されています。天文学と物理学，化学，地球科学です。無機化合物に起こる現象を体系立てて観察する学問であり，生命科学の領域である自然界の観察とは大きく異なります。

　物理科学に今求められているのは科学技術への応用であり，その研究は 2 つの領域に集中しています。それはすでに発見されている物質の新しい特性を見つけ，より良質な，あるいは廉価な製品を作ること，また，これまで存在しなかった新しい物質を作り出すことです。

　一世紀前まで，物理科学者はニュートンの示したモデルの中で研究してきました。不確実なものではなく確実なものを追究し，すっきりとした数式で表現，問題を解決するという姿勢です。ところがこの 20 世紀の中で，世界（宇宙）はそのような構造にはなっておらず，現実世界には混沌や不安定，不確実性が存在することに科学者は気がついてきました。そしてこの現実に対応できる術を築いていきました。その代表的なものが量子力学です。

　理論科学の分野では，量子力学と重力の枠組みを示すアインシュタインの一般相対性理論とを統一した理論を打ち立てようとしています。この困難な試みの中から，超弦理論のような新しく興味深い考え方が現れました。現在では計算機科学により，車の流れや人の動き，物質をさまざまに組み合わせた際に生じるふるまいと相互作用，バクテリアの細胞の発達までもシミュレーションで表現可能になりつつあります。その結果，新しい画期的な技術が生まれるかもしれません。

　対象が液体であれ，大気，海，温室ガスで構成される気候のモデルであれ，物理科学者は，諸現象は単一の要素ではなく，複数の要素の相互作用で生ずるのだということを認識するようになりました。そのため今強調されているのは学際的物理科学，物理科学を構成する諸分野の最先端の成果を取り入れた研究です。

Chapter 19

Describing actions and events in sequence

If you describe actions **in chronological order**, the reader will understand what is going on and will not **feel disoriented**. When describing an order of events in English, it is critical to use different words that accurately describe the sequence of events. The more words you know in English to express yourself, the easier it will be to express yourself clearly. For presentations, sequence diagrams are quite useful for explaining the order of actions and events. They can typically communicate fairly complex sequences with conditions, branches, and loops.

in chronological order 時系列で
feel disoriented 迷う，戸惑う

19.1 Sequence

Sequence refers to the order of steps in which a procedure is carried out. An introductory statement or sentence lets the reader know something general about the subject to be discussed in the paragraph. Then the explanation should be in a **temporal sequence** which means using words that indicate a time connection. Temporal sequential ordering thus organizes the different steps of logic, thought, and action by time.

temporal sequence 時系列
temporal 時間の

When *yet* is used with the present perfect or present perfect continuous, it means at any time **up to** the present. It is used mostly in questions and negative sentences.

up to … ～まで

Example:
· The **plumber's** career was entirely based on physics, *yet* he did not even know what physics is.

plumber 配管工

19.2 Language expressing sequence

Time sequencing markers:

First, secondly, next, after, following, once, when, as soon as, then, before, now, once, afterwards, while, during

19.2.1 Sequence of tenses

When writing a complex sentence (a sentence that includes several clauses), the main idea is placed inside the **main clause** rather than the **subordinate** or dependent **clause**. If the verb of the main clause is in the present tense, the tense of the verb of the subordinate clause can take any logical form.

main clause 主節
subordinate clause 従属節

Examples:
· **Liquefaction** occurs when the density of sand is less than a certain value.
· Liquefaction occurs if the density of sand was less than a certain value.

liquefaction 液化

The subordinate clause must be in the **past** or **past perfect tense** if the main clause is in the past or past perfect tense.

past tense 過去時制
past perfect tense 過去完了時制

(1) To show priority, the past perfect tense has to be used in the subordinate clause.

Example:
· They reported that all the needed experiments had already been carried out.

simultaneity [ˌsaɪmltəˈniːəti] 同時性

(2) To show **simultaneity**, the simple past or past **continuous** tense should be used in the

continuous 連続した

subordinate clause.

Examples:
- The report stated that they were wrong in their conclusions.
- The professor asked whether we were preparing for the test.

(3) To show future actions, *should* or *would* to indicate the future is used with the past tense.

Example:
- It was uncertain whether he would be able to attend the conference in time.

| 19.2.2 | **Using participles to indicate sequence** |

participle 分詞

To show sequence, the earlier event can be written using the perfect participle.

Example:
- Having noted that the units are **consistent**, they were temporarily removed from the equation.

consistent 一貫している
⇒W11

Text: Innovations

Innovations introduce something new or make a change to any present product, idea, or field. There is a slight difference between inventions and innovations. Inventions refer to acts of designing, discovering, or creating a new procedure, product, device, or technique; innovations modify or update an existing product, device, approach, or system so that its value is **enhanced**.

There is, <u>without a doubt</u>, no proven **formula** for success, particularly when it comes to innovation. President John F. Kennedy's **audacious** goal of "going to the moon in this decade" in 1962 **inspired** the American people to <u>**unprecedented**</u> levels of innovation in a variety of fields, not just space exploration. A far-reaching vision can thus be an **enticing catalyst**, as long as it is realistic enough to motivate action today.

Innovations in products or processes normally observe a route from a laboratory (lab) idea, through pilot or prototype manufacturing and production start-up, to **full-scale** production and marketplace introduction. Thus research is basically the first stage in the development process. Research activities may be aimed at achieving either specific or general objectives. It is definitely beneficial to combine high-level goals with estimates of the value that the innovation may generate. Because no one knows where valuable innovations will emerge, and searching everywhere is impractical, researchers must initially establish some **boundary conditions** for the fields they wish to investigate. The process of identifying and defining these spaces can range from intuitive future visions to **meticulous** strategic analyses.

Exceptional creativity is just one of the approaches that leads to innovations. **Methodically** and **systematically scrutinizing** three areas can yield insights into potential innovations: a valuable problem to solve, a technology that enables a solution, and a business model that generates money from it. Probably every successful innovation occurs when these three elements intersect.

Innovation is the type of change our society understands best. We must, <u>however</u>, be careful so that these innovations themselves are not used to negatively impact their users. Innovation should be about doing new things to solve problems and improve the lives of people all around the world.

- *Innovate* – verb-to begin or introduce (something new) for the first time
- *Innovation* – noun - a new technique or idea
- *Innovative* – adjective - being or producing something new
- *Innovator* – noun - someone who helps to open up a new line of research or technology or art

; ⇒G16.2.1
enhance 高める
without a doubt ⇒G16.3
formula 公式
audacious 大胆な
inspire 促す，鼓舞する（心 (spirit) を中 (in) から入れる）
unprecedented かつてない，前代未聞の ⇒W15
enticing (形容詞)魅力的な
catalyst 誘因
full-scale 本格的な

initially まず
boundary condition 境界条件
meticulous 綿密な
methodically 整然と
systematically 整然と
scrutinize 精査する

however ⇒G16.3

innovative [ˈɪnəveɪtɪv] 革新的

Chapter 19

Table 19.1 Technial vocabulary related to R&D

Words	Definition
pure basic research	fundamental theoretical or experimental investigation to enhance scientific understanding; **instant practical utility** not being a direct goal
applied research	studies directed toward using **expertise** obtained via basic research to make things or to create conditions **in such a way as to** serve a real purpose.
experimentation	the procedure of tests and trials to **probe** what occurs under different conditions
innovation	a new idea, device or method
pilot study	a small-scale experiment or series of observations carried out to determine <u>how and whether to</u> launch a larger project
product development	changing and improving a product to attain the highest quality possible
analysis	the study of the components and their interactions with each other

instant practical utility 効用
expertise［ˌekspɜːrˈtiːz］専門知識
in such a way as to … ～ の方法で
probe 調査する
<u>how and whether to</u> ⇒ⓒ12.6

Exercises

Understanding the text
Q1 Find the words or expressions that indicate a sequence in action in the Text: Innovations.

Grammar point
Q2 Combine the two sentences into one sentence using the word or phrase in the bracket to indicate sequence of actions.

i. They **fire** the RCS thrusters. They turn the **orbiter** tail first. (then)

fire 発射する
orbiter 宇宙船，軌道船
harmonic motion 調和運動

ii. The block moves in a straight line in simple **harmonic motion**. It has been either pulled or pushed away from its original position before being released. (once)

iii. The **displacement** is zero. The acceleration is also zero. (when)

displacement 変位

iv. The circuit is finished. The charge starts to flow. (as soon as)

v. The switch is closed. Current flows through the resistance. (after)

Writing
Q3 Write a paragraph about an interesting feature of the Japanese education system. Use at least four temporal sequencers, i.e. time markers.

第19章　行動や事象を時系列で述べる

　行動を時系列で提示すれば，読者は今起きていることがわかり，事象の前後関係で混乱することはありません。時系列で提示するには，時系列を正確に表現する語を用いなければなりません。その語彙を多く使うことができれば，より正確に表現することが可能になります。プレゼンテーションの際には，シーケンス図（時系列を示す図）は行動や事象の順序を説明するのにとても便利です。シーケンス図は通常，条件や分岐，ループを含むかなり複雑な時系列を伝えることができます。

19.1　時系列

　時系列（シーケンス）は，一つの手順が実行されるステップの順序を指します。冒頭の文によって読者は，そのパラグラフで論じられるテーマについて一般的なことを知ることができます。次に，時間の順序，すなわち時間的なつながりを示す単語を使って説明します。このように時間的順序付けは，論理と思考，行動のさまざまなステップを時間によって整理します。

yet が現在完了や現在完了進行形で用いられるとき，「現在に至るまで」を意味します。多くは疑問または否定文に現れます。

例：
- The plumber's career was entirely based on physics, **yet** he did not even know what physics is.（配管工の仕事は物理学の知識に基づいていた。しかしこの配管工は物理学とは何かすら知らなかった。）

19.2　時系列を表す表現

タイムシーケンスマーカー：

first, secondly, next, after, following, once, when, as soon as, then, before, now, once, afterwards, while, during

19.2.1　時制の時系列

2個以上の節からなる複文を書くとき，主要な考えは，従属節ではなく主節に置かれます。主節の動詞が現在形であれば，従属節の動詞の時制は何でも構いません。

例：
- Liquefaction occurs **when** the density of sand **is** less than a certain value.（液化が生じるのは，砂の濃度が一定の値を下回るときだ。）
- Liquefaction occurs **if** the density of sand **was** less than a certain value.（液化が生じるのは，砂の濃度が一定の値を下回ったときだ。）

主節の動詞が過去形あるいは過去完了形であれば，従属節の動詞も同じ形をとります。

(1)　先行した時間を示すには，従属節の動詞を過去完了形にする

例：
- They **reported** that all the needed experiments **had** already been carried out.（報告では，必要な実験はすべて実施したとのことであった。）

(2)　同時間を示すには，従属節の動詞を過去形ないし過去進行形にする.

例：
- The report stated that they **were** wrong in their conclusions.（報告では，結論に誤りがあると述べられた。）
- The professor asked whether we **were** preparing for the test.（教授は私たちに試験の準備をしているのかと尋ねた。）

(3)　未来の行動を示すには，should や would といった未来を示すことばが過去形で使われる

例：
- It was uncertain whether he **would** be able to attend the conference in time.（彼が会議に間に合うのかはっきりしなかった。）

19.2.2　分詞による時系列の表現

時間の連続性を示すのに，先行する事象を完了分詞形で表します。

例：
- **Having noted** that the units are consistent, they **were** temporarily **removed** from the equation.（単位が一定であることがわかったので，とりあえず方程式から除外された。）

テキスト：イノベーション

　イノベーションにより，現在の製品やアイデア，研究分野に新しいものが加えられ，変化が生まれます。発明とイノベーションにはいくらか違いがあります。発明は，新しい製品や手法，デバイス，技術のデザイン，発見あるいは創造の行為であるのに対し，イノベーションは，今ある製品やデバイス，研究法，システムに手を加えてアップデートし，それらの価値を高めます。

　もちろん，ものごとを前進させるのに定まった方程式は存在しません。イノベーションについてはとりわけそうです。1962年にアメリカの故ケネディ大統領が行った「この10年内に月へ」という大胆な呼びかけに強く触発され，アメリカでは宇宙技術のみならずさまざま

な分野でイノベーションが起こりました。このようにはるか先を見通した提案は，次代を開く力強い牽引力となりえます。ただしそれは，人の意欲をかき立てる現実的（実現可能）なものでなければなりません。

製品と製品化に至るプロセスとのイノベーションは，研究室での着想，プロトタイプ作製と製品化着手，本格的な製品の生産と市場販売の流れの中で生じます。したがって研究は，製品化への道の第一歩となります。研究活動は，個別あるいは一般性をもつ目標の達成を目指してなされます。高く掲げた理想とイノベーションのもたらす新たな価値とをつなぐのは確かに有益です。価値あるイノベーションがどの段階で生じるのかは予測できず，また個々の段階のすべてをチェックするのは非現実的ですので，研究者はまず，研究対象を絞り込む必要があります。対象を確定し限定する作業は，直感的な未来像から製品化への入念な戦略作りまで多岐のものを含みます。

ずば抜けた創造性以外にもイノベーションを生む道，手立てはあります。次の3つを整然と精査すれば新たなイノベーションの手がかりをみつけることができます。重要な問題の設定，その課題の解決を可能にする技術，市場につながるビジネスモデル，です。すぐれたイノベーションは，すべてこれらの3つの要素が相まって生まれているといってよいと思います。

イノベーションという変革は，今の社会でよく理解されています。しかしここで注意しなければならないことは，その使い方を誤らない，ということです。イノベーションは新しいものの提示により問題を解決し，世界中の人々のくらしをよくするものでなければなりません。

Innovate – 動詞 – 新しいことを初めて行う，あるいは導入する
Innovation – 名詞 – 新しい技術やアイデア
Innovative – 形容詞 – なにか新しいものである，あるいはなにか新しいものを生み出す
Innovator – 名詞 – 新しい研究，技術，芸術の創造に貢献する人

表 19.1 研究開発（R&D）に関係する用語

語	定義
pure basic research （純粋な基礎研究）	科学的理解を深めるための基礎的な理論的または実験的研究。すぐに実用化されることを直接的な目標とはしない
applied research （応用研究）	基礎研究によって得た専門知識により，ものを作る，または実際の目的に役立つ条件を整えることを目的とした研究
experimentation （実験）	異なる条件下で起こることを探るための試験や試行の手順
innovation （イノベーション）	新しいアイデアや装置，方法
pilot study （パイロット・スタディ）	大規模なプロジェクトをどのように立ち上げるか，あるいは立ち上げるかどうかを決定するために実施される小規模な実験または一連の観察またプロジェクトを開始するかを検証する
product development （製品開発）	可能な限り最高の品質を実現できるよう製品を変更し改良すること
analysis （分析）	構成要素の，また構成要素間の相互作用の研究

Chapter 20

Giving a definition

A paper's meaning can be **strengthened** by defining specific words which are relevant to the subject matter. Adding definitions to a report or paper effectively is a process that includes deciding which words to define, rephrasing definitions, citing the source of the definition, defining the term in the context of the paper, and keeping the definition brief.

strengthen 強化する

20.1 Providing a definition

In order for your research to be accepted, it must be shared within the scientific community. When making a hypothesis or other **assertion**, scientists need to make certain that they are unambiguously understood by other researchers. Sometimes scientific disagreements can occur when people misunderstand each other. To avoid this, definitions which help to clarify certain words and concepts are necessary. At other times, a definition is needed due to the fact that a particular term is being used in a particular way, e.g., the case of *work* and *energy* when used in the context of physics. Communication between researchers is dependent on precise definitions of ideas, concepts, materials and techniques.

assertion 主張

In the Introduction section of a research paper or report, it is sometimes necessary to provide definitions to explain important technical words to the reader. In the case where a word needs to be found which is better or clearer than the word under consideration, a dictionary can be used to provide the meaning or the **thesaurus** to provide easily understood **synonyms**.

thesaurus 類義語辞典
synonym 同義語

A sentence providing a definition is generally divided into four elements: (1) the noun itself, (2) a *to be* verb, (3) a **generic noun**, and then (4) the main function.

generic noun 一般名詞

Example:
· A **transmitter** is *a device* (generic noun) that generates signals.

transmitter 送信機

More complex definitions may have additional information before or after the generic noun.

Example:
· A **thermocouple** is *a device*, consisting of two different metals **welded** together at each end, which is used <u>to accurately measure</u> temperatures.

thermocouple 熱電体
weld 溶接する
<u>to accurately measure</u>
⇒ Ⓖ12.8

20.1.1 Defining relative clauses

A **relative clause** may be used to offer additional information about a noun, i.e., provide a definition of a person, place or thing. A defining relative clause gives the information that directly identifies what is being talked about.

relative clause 関係詞節

Examples:
· A portal is a website which serves as an entry point to other websites, frequently by hosting or providing access to a search engine.
· An inventor is a person who **comes up with** an idea for something new.
· An alarm is a device that is used to warn of danger by means of a sound or signal.

come up with 思いつく

> **Non-defining relative clauses:** Relative clauses, which add more information about nouns, but do not provide a definition of the noun being talked about are termed non-defining relative clauses. They are introduced and followed by a comma (if they do not end the sentence).
>
> Examples:
> · This is the main transmitter, from which all the signals are sent.

> · The main transmitter, from which all the signals are sent, is located in Tokyo.

20.1.2　Using infinitives or prepositions

Definitions can sometimes be written with the help of an infinitive or **prepositional phrase**.

prepositional phrase 前置詞句

Examples:
· Energy is the capacity to do work.
· Energy is the capacity for work.

20.1.3　Using modals

When using the modal verb of possibility *may* to construct a definition, it gives the impression that there may be other ways of defining.

Example:
· For a closed thermodynamic system, entropy may indicate a **measure** of the amount of thermal energy not available to do work.

measure 尺度

20.1.4　Using adjectives and synonyms

When writing definitions, adjectives can be used to describe the type of object to which the term refers. In addition, synonyms can also be used to give a better understanding to the reader of the technical vocabulary they are encountering.

Examples:
· Potential energy is stored energy.
· The algorithm utilizes a heuristic or rule-of-thumb approach.

20.2　Tense and article

Definitions are written using the form of A is B. The definite article, the, is not used before the defined word since definitions are general statements.

Example:
· An **insulator** is a substance that does <u>not transmit heat, sound, or</u> electricity.

insulator 絶縁体
not transmit heat, sound, or
⇒ Ｇ16.7

20.3　Language patterns indicating definition

Sometimes it is necessary to provide a further explanation of a difficult concept or technical word. In that case, the following phrases are a good way to begin the definition:
What is meant, For example, There are several ways to look at etc.

Examples:
· Keep in mind what is meant: power is the rate at which a force is applied to do work.
· The tires of a car, <u>for example</u>, deform easily under load.
· There are several ways in which virtual and real images are created by **reflection** and **refraction**.

for example ⇒ Ｇ16.3
deform ⇒ Ｗ8
reflection 反射
refraction 屈折

Some words, as well as patterns of writing, also help to signal that a definition will be provided at this point of the text. The following sentences provide examples to illustrate how the meaning of a *word* is positioned:
· *Word, ... they are* *meaning*: The lenses are *compound* lenses, which means they are *made of several components*, with interfaces that are rarely perfectly **spherical**.
· *Word (meaning)*: Silicon and germanium are *good semiconductors (materials that are neither good conductors of electricity nor good insulators but have properties of elec-*

compound 複合
spherical 球状の

trical conductivity somewhere between the two).

- *Abbreviation* - i.e. (that is): The oscillator can generate *550 nm* (i.e., visible) electromagnetic waves.

abbreviation 略語（ab-（＝ad- toward）（brev-（＝brief 短い））

- *Word*. It is *meaning*: *The primary rainbow* is the type commonly seen in regions where rainbows appear. It is *caused by light refracting and reflecting inside the water drops*.

- *Meaning ... is/are ... called a word*: *This type of image exists only within the brain* and is called a *virtual image*.

- *Meaning ... is/are ... known as word*: *A style of painting with **striking** illusions of color* is known as *pointillism*.

striking 印象的な

- *Word, that is, meaning*: *The measurement requires **resolving** lines with close wavelengths*, that is, *making the lines distinguishable*.

resolve 分解する

- *Meaning means word*: *Locating an object* means *determining its position in relation to some **reference point***.

reference point 基準点

- *Meaning, or (a) word*: *The position of a particle **on an axis** determines the position of the particle with respect to *the origin*, or *zero point, of the axis*.

on an axis 軸上の

- *Full expression, shortened to **abbreviation***: *Foods that contain genetically altered **ingredients** are known as **genetically modified food***, shortened to *GM food*.

ingredient 成分
genetically modified food 遺伝子組み換え食品

Chapter 20

20.4   Writing a good definition

A good definition ensures that your readers always understand clearly what you are trying to express in your writing.

(1) The definition should always include the class of object for which the explanation is being provided. To **help with** the definition, examples, comparisons and details may be given, but it is **imperative** that the class not be excluded.

help with 助ける（＝ help you with）
imperative 必須だ

Example:
- Calculus is a branch of mathematics.

(2) Precise definitions of **technical terms** are important. Just saying "Carbon is a chemical element" is not enough; more specific information is needed.

technical term 専門用語

(3) The definition should never be more difficult than the term that is being explained.

(4) Sometimes definitions need to say what the term is not and is thus called a negative definition. If such a definition is needed, it should always be followed by a statement expressing what the term actually <u>does mean</u>.

does mean ⇒ G 16.6

(5) The definition must express or deal with facts without being **distorted** by personal feelings, **prejudices**, or interpretations.

distort ゆがめる
prejudice 片寄った見方（pre-（前もっての）＋jud-（＝ judge 判断））⇒ W 6

20.5   Checking the quality of a definition

A good way of checking the completeness of a definition is by **reversing** it. For example, if it is written

reverse 入れ替え

"A cyclotron is an **apparatus**.",

apparatus [ˌæpəˈrætəs] 装置

then reversing it, we get the sentence

"An apparatus is a cyclotron."

which does not make sense. However, if it is written

"A cyclotron is an apparatus which generates charged particles and accelerates them spirally outward.",

then the reverse would be

"An apparatus which generates charged particles and accelerates them **spirally** outward is a cyclotron.",

spirally 螺旋状に

which is perfectly **comprehensible**.

comprehensible 理解でき
る

Text: Young scientists

Scientists try to make our everyday lives easier and build a better world. The main feature of young scientists is to recognize and understand specific issues so that they can work to find the remedy to those problems. The vast majority of young scientists possess confidence, ambition, and determination, but to fully achieve their career goals, some other **issues** also need to **be addressed**.

issue 課題
be addressed 問題に取り組
む

Professional success is a matter of practical experience, judgment, **foresight** and luck. Before young scientists begin on their research, they should try to find out the underlying objectives their project is trying to accomplish, **i.e.**, succeed in doing. The success of the research project will be measured by how well these **aspirations** are met. When a young scientist is trying to achieve success in his/her research, the **preliminary** setting of realistic goals is crucial, **that is**, essential or vitally important for the accomplishment of his/her tasks.

foresight 先見の明

aspiration 強い願望
are met ⇒ⓖ16.5
preliminary 予備的な

Young scientists must understand that it is not about just learning one set of skills; they need to realize that it is **imperative** to continuously learn over the course of their whole lives. No matter how good or original or appealing an idea is, **it is only as valuable as it is practical**. In order to be productive, they therefore need to plan their work and **execute** it **accordingly**. Whatever they are working on, they need to create a routine. Times for specific activities should be scheduled, and **rigorously adhered to**. In theory, everyone understands that preparation can make or break an important task.

; ⇒ⓖ16.2.1
imperative 必須である
it is only as valuable as it is
practical 実用的であって
こそ価値がある（直訳は
「実用的と価値の程度は同
じ」）
execute 実行する
accordingly それに従って
⇒ⓖ5.6
rigorously 厳格に
adhered to 守る ⇒ⓦ17

Young scientists also need to be able to speak clearly and naturally. Nowadays, since the international language is English, lack of ability in speaking English clearly is often perceived as a lack of ability to think clearly about science. This is of course wrong, but young scientists need to increase their English communication skills of speaking and writing.

Finally, always acknowledging other's contributions enhances effective communication and strengthens relationships between research members. How you show respect to others, how you share ideas, how you acknowledge other's contributions, and even how you talk about your own efforts will affect the success or failure of your future career.

Exercises

Understanding the text

Q1 Fill in the gaps with the words **denoting** a specialist in that particular field.

denote 示す，指す

For example: science scientist

chemistry _____   physics _____

zoology _____   engineering _____

genetics _____

Q2 Match the definitions (1–4) on the left to the words (a–d) on the right.

i. ___ to learn or understand something completely   (a) to develop

ii. ___ to think of or produce a new idea, product,
etc. and make it successful   (b) to master

iii. ___ a careful study of a subject   (c) science

iv. ___ knowledge about the structure and behavior of the natural and physical world　　(d) research

Grammar point

Q3 Combine the two sentences into one sentence using *which*, *who* or *that* to provide a definition.

　i. A hovercraft is a vehicle. It travels across land or water.

　ii. GPS or Global Positioning System is an electronic system. It uses a network of satellites to indicate the position of a vehicle, ship, person, etc.

　iii. An **ammeter** is an instrument. It measures the strength of an electric current in terms of amperes.

ammeter 電流計

　iv. Torque is a **twisting** force. It generally causes something to rotate around an axis or another point.

twist ひねる

Writing

Q4 Use the information given below to write sentences expressing a definition.

　i. **Acoustics** / science / sound

　ii. **Catalyst / substance** / speeds up the rate of a chemical reaction.

acoustics 音響学
catalyst 触媒
substance 物質

Chapter 20

第 20 章　定義づけ

　主題に関する特定の語を定義しておくと，論文の価値は高まります。レポートや論文に定義を効果的に加えるには，定義する語の選択，定義の言い換え，定義の出典の提示，文脈に応じた定義，簡潔な定義を心がけてください。

20.1　定義の提示

　研究が受け入れられるためには，その研究が科学界で共有されなければなりません。仮説を立てたり他の主張を行ったりする場合には，科学者は他の研究者に明確に理解される必要があります。科学的な意見の相違は互いの誤解から生じることがありえます。それを避けるためには，特定の言葉や概念を明確にするための定義が必要です。また，特定の用語が特定の方法で使用されているために定義が必要になることもあります。たとえば，物理学の文脈で使用される仕事とエネルギーなどです。研究者間のコミュニケーションは，アイデアや概念，材料，技術の正確な定義に依存しているのです。

　学術論文やレポートの序論では，重要な専門用語を読者に説明するために定義の提示が必要な場合があります。使用を検討中の単語よりも，優れた，あるいは明確な単語を見つける必要がある場合，辞書を使って意味を調べたり，類義語辞典を使って簡単に理解できる同義語を調べたりすることができます。

　定義を提供する文は，一般的に4つの要素に分けられます：(1) 名詞そのもの，(2) to be 動詞，(3) 一般名詞，そして (4) 主な機能です。

例：

・A transmitter is *a device*（一般名詞）that generates signals.（送信機とは信号を発する装置である。）

より複雑な定義は「a + 一般名詞」の名詞の前，または後ろにさらに情報を加えます。

例：

・A thermocouple is *a device*, consisting of two different metals welded together at each end, which is used to accurately measure temperatures.（熱電体は溶接された一対の異なる金属で，温度を正確に測るのに用いられる装置である。）

20.1.1　定義に用いられる関係詞節

　関係詞節を用いて名詞についてさらなる情報を与えることがあります。すなわち，人，場所，ものを定義します。定義に用いられる関係詞節は，何が話題になっているかを直接的に特定する情報を与えます。

例：

- A portal is a website **which** serves as an entry point to other websites, often by being or providing access to a search engine. (ポータルとは，検索エンジンあるいは検索エンジンへのアクセスを提供することで，他のウェブサイトに入る入口のウェブサイトのことをいう。)
- An inventor is a person **who** comes up with an idea for something new. (発明家とは，何か新しいアイデアを思いつく人のことである。)
- An alarm is a device **that** is used to warn of danger by means of a sound or signal. (アラームとは，音や信号で危険を知らせる装置である。)

> **定義に用いられない関係詞節**：名詞に追加の情報を与えるが定義はしない関係詞節を定義に用いられない関係詞節（non-defining relative clauses）と呼びます。この関係詞の前に「,」がきます。
>
> 例：
> - This is the main transmitter, from which all the signals are sent. (これが主たる送信機である。そこからすべての信号が送られる。)
> - The main transmitter, from which all the signals are sent, is located in Tokyo.

20.1.2　不定詞あるいは前置詞の使用

不定詞句あるいは前置詞句を用いた定義があります。

例：

- Energy is the capacity **to do** work. (エネルギーとは，作用する能力のことをいう。)
- Energy is the capacity **for work**. (同上)

20.1.3　助動詞の使用

可能性を表す助動詞の may を用いて定義すると，他の定義の仕方もありうる，との印象を与えます。

例：

- For a closed thermodynamic system, entropy **may** indicate a measure of the amount of thermal energy not available to do work. (閉じた熱力学システムでエントロピーは仕事に使えない熱エネルギーの量の尺度を示す可能性がある。)

20.1.4　形容詞と同義語の使用

定義を書く際には，形容詞を用いて定義する対象を記述することがきます。さらに，同義語を使用することで，読者が目にする専門用語をよりよく理解することができます。

例：

- Potential energy is stored energy. (ポテンシャルエネルギーとは，内蔵されたエネルギーのことをいう。)
- The algorithm utilizes a heuristic or rule-of-thumb approach. (このアルゴリズムは，発見的あるいは経験則に基づく手法をとっている。)

20.2　時制と冠詞

定義は「A is B」の形で書かれます。定義は一般的な記述であるため，定冠詞の the は通常，定義には使用されません。

例：

- An insulator **is** a substance that does not transmit heat, sound, or electricity. (絶縁体は，熱や音，電気を伝達しない物質だ。)

20.3　定義を示すパターン

難しい概念や専門用語を定義するには，さらに詳しい説明が必要となる場合があります。その場合，以下の語句から定義を始めるとよいでしょう。

What is meant, For example, There are several ways to look at などです。

例：

- Keep in mind **what is meant**: Power is the rate at which a force is applied to do work. (次の定義を覚えておきましょう。仕事率（もしくはパワー）とは，仕事をするために力を加える速度のことである。)

- The tires of a car, **for example**, deform easily under load. (たとえば，車のタイヤは負荷をかけると容易に変形する。)
- **There are several ways** in which virtual and real images are formed by reflection and refraction. (反射と屈折により仮想と現実のイメージを形成する方法がいくつかある。)

定義がテキストのこの時点でなされることを知らせる語や表現があります。以下の文は，語の意味が文のどの場所に置かれるかを示しています。

- *Word, ... they are meaning*: The lenses are *compound lenses*, which means they are *made of several components*, with interfaces that are rarely being perfectly spherical. (これらのレンズは複合レンズである。いくつかの部分からなり，それらの接点が完全な球形になることはほとんどない。)
- *Word (meaning)*: Silicon and germanium are *good semiconductors* (*materials that are neither good conductors of electricity nor good insulators but have properties of electrical conductivity somewhere between the two*). (シリコンとゲルマニウムはすぐれた半導体である（半導体は電気をよく通すとかすぐれた絶縁体というわけではないが，その中間の電気伝導率をもつ材料である。))
- *Abbreviation- i.e. (that is)*: The oscillator can generate *550 nm* (i.e., *visible*) electromagnetic waves. (略語表現 i.e. = that is（すなわち）：オシレーター（振動子）は 550 nm（＝可視）電磁波を生み出すことができる。)
- *Word. It is meaning*: The primary rainbow is the most common type seen in regions where rainbows appear. It is *caused by light reflacting reflecting inside water drops*. (主虹は虹の現れるところに最もよく見られる虹である。これは水滴を反射する光により発生する。)
- *Meaning ... is/are ... called a word*: This type of image exists only within the brain and is called *a virtual image*. (このタイプのイメージは脳内にのみ存在し，仮想イメージと呼ばれる。)
- *Meaning ... is/are ... known as word*: A style of painting with striking illusions of color is known as *pointillism*. (印象的な色の幻想で描く手法は点描画法として知られている。)
- *Word, that is, meaning*: The measurement requires resolving lines with close wavelengths, that is, *making the lines distinguishable*. (この測定では波長が近い線を分解する，つまり線を区別できるようにする必要がある。)
- *Meaning means word*: Locating an object means *determining its position in relation to some reference point*. (位置を定めるとは，基準点との関係で位置を確定することだ。)
- *Meaning, or (a) word*: The position of a particle on an axis determines the position of the particle with respect to *the origin*, or *zero point, of the axis*. (軸上の粒子の位置は，軸の原点またはゼロ点に対する粒子の位置を決定する。)
- *Full expression, shortened to abbreviation*: Foods that contain genetically altered ingredients are known as *genetically modified food*, shortened to *GM food*. (遺伝子組み換え成分を含む食品は遺伝子組み換え食品，短縮して GM 食品として知られている。)

20.4　すぐれた定義の書き方

すぐれた定義を与えると，著者が言おうとしていることを読者はいつも明確に理解します。

(1)　定義するにはまず，説明の対象となるものの分類をつねに含めなければなりません。定義を助けるため，例示，比較，具体的記述を記すことができますが，その分類が除外されないようにすることが重要です。

　　　例：
　　　・Calculus is a **branch of** mathematics. (微積分学は数学の一分野である。)

(2)　専門用語の正確な定義が重要です。"Carbon is a chemical element（炭素は化学元素である。）." というだけでは十分ではありません。もっと具体的な情報が必要です。

(3)　定義される語より難しい語を使って定義してはいけません。

(4)　定義には，その用語が何を意味しないかを示す必要がある場合があり，それを否定的定義と呼びます。そのような定義が必要な場合は，つねにその後にその用語が実際には何を意味するのかを表す文を続けるべきです。

(5)　定義に用いられる事実は個人的感情や先入観，解釈によってゆがめられていてはなりません。

20.5　定義の質の確認

定義が適切であるかどうかを確認良い方法は，その定義を逆にすることです。たとえば，

"A cyclotron is an apparatus. (サイクロトロンは装置である)"

と書かれた場合，それを逆にすると

"An apparatus is a cyclotron."

という意味不明な文になります。しかし，

"A cyclotron is an apparatus which generates charged particles and accelerates them spirally outward.（サイクロトロンは荷電粒子を発生させ，それを螺旋状に外側に加速させる装置である。）"

と書かれていれば，その逆は

"An apparatus which generates charged particles and accelerates them spirally outward is a cyclotron.（荷電粒子を発生させ，それを螺旋状に外側に加速させる装置はサイクロトロンである。）"

となり，完全に理解できます。

テキスト：若い科学者

　科学研究者は暮らしと世界をよりよいものにしようと努力しています。若い研究者は特定の課題を認識，理解し，それらの問題の解決に取り組みます。多くの若い研究者には自信と大志，固い決意があります。しかし，目標を十分に実現するには，他の課題にも目を向ける必要があります。

　科学者として成功を収めるには，実践と判断力，先々への見通し，幸運が必要です。研究に取りかかる前に，若い研究者はこれから取り組む具体的な研究の底にある，もっと大きな枠組み（究極的な研究目標）の把握に努めてほしいと思います。研究の成功の度合いは，研究者の意欲がどの程度報われたかによって計られます。若い研究者が研究で成果を上げようとする時，前もっての達成可能な目標設定は極めて重要です。

　若い科学者は今行っている研究に必要な一連の技術を身につけるだけでなく，研究者として活動する間はずっと学び続けなければならないことを認識しなければなりません。一つのアイデアがどれほどすばらしく魅力的であっても，それが真に価値のあるものであるためには，そのアイデアは実用的であってこそ価値があります。したがって，生産的な研究を行うためには，彼らは研究の計画を立て，それに沿って実行しなければなりません。日々の研究のルーティーンを作り，具体的な活動のスケジュールを作成し，それを徹底的に守るのです。入念な準備が成功を可能にすると，だれでも頭の中ではわかっているのです。

　また，若い科学者は明確にかつ自然に話せるよう努めてください。世界の共通言語が英語ですので，英語を話す力が劣っていると，科学について明確に考える力が欠けているのでは，と誤解されかねません。これはもちろん間違った考え方ですが，若手の科学者は英語で話し，書く技術の鍛錬を怠りなくする必要があります。

　最後に，他の研究者の貢献への感謝の気持ちをもてば，効果的なコミュニケーション力が高まり，研究者間の人間関係が良好になります。他者への尊敬の心を失わずに，研究テーマについて意見を日常的に交換し，他の研究を正しく評価して，自分の努力もきちんと口にする。こうすることで，研究者，技術者としての将来は明るいものになります。

Chapter 21

Describing use or function

use 用途
function 機能

Use or function of something refers to what an object or research project actually does in the real world. So, to describe a function of a study in detail includes making suggestions, criticizing, refusing, agreeing and disagreeing, inquiring, discussing the past, and offering advice. Certainly, we have to describe the purpose and usefulness of the study under consideration.

21.1 Describing use or function

The following structures are used to describe the purpose of a function or **device**.

device 装置

(1) (*in order*) *to* + infinitive

infinitive 不定詞

Examples:
- A voltmeter is used to find the potential difference between any two points in a **circuit**.
- In order to monitor solar flares, a space **probe** needs to be positioned along a line that extends directly toward the Sun.

circuit 回路
probe 探査機

(2) *for* + verb + *-ing*

Example:
- Thomson's apparatus is used for measuring the **ratio of** mass **to** charge for an electron.

ratio of A to B
A の B に対する比

(3) *act as* + noun

Example:
- Immediately after the switch is closed, the inductors act as broken wires.

21.2 Aim or objective

aim 目的
objective 目的

The purpose of an action can be addressed in writing or speaking by the following ways:

(1) *Aim/purpose/objective/goal*:

Examples:
- We need to aim for precision in science.
- One of the goals of physics is to **document** observations about our world.
- The primary objective is to obtain a nice correlation between its systems.

document 記録する

(2) Combined with *if/ whether*:

Example:
- A related aim/purpose/objective/goal is to determine whether a force is affected by gravitation.

(3) *That/which* + present simple:

Examples:
- An example of an emf device is a thermopile that provides onboard electrical power in some spacecraft.
- The aim of the experiment, which investigates the effects of temperature, is to discern patterns and draw conclusions based on **empirical** data.

empirical 経験的な

Text: The importance of conferences

Networking is the exchange of information and ideas among people who share a common profession or another interest. Networking helps a professional to stay up to date on current events in the field, as well as providing occasions to assist people in finding future employment opportunities. Good networks are built on good relationships. Good communication is an essential component of all relationships. The act of communicating not only helps you to meet your needs, but it also allows you to share positive feelings, thoughts, and suggestions with your peers. Professional conferences are thus a necessary fact of working life. Conferences are a very important way for researchers to stay connected to others. A good conference can have the capacity to bring a scientist **out of a rut**, regardless of his/her career stage.

> **out of a rut** 退屈させない

It is not possible for people to remember every slide and every word of a conference presentation. So, scientists should explore all **avenues** for keeping others intellectually engaged and thus benefiting from their advice. Although professional conferences are **geared** primarily **toward** professors, students can also benefit from scientific conferences by contributing and presenting their own data as well as learning about advances in their subject. They should always write down the most exciting and important ideas. Meeting other people is essential for enabling collaboration and creativity. The main benefits of conferences are thus feedback and collegial support. They are also very helpful for learning about new topics and publications. For many, in addition, they provide fun opportunities to meet people.

> **avenue** 道，手立て
>
> **geared toward ...** 〜を対象とする
>
> **collegial** 同僚による

Exercises

Understanding the text

Q1 Match the actions with their purposes.

Action (i〜iv)	Purpose (a〜d)
i. You may experiment	(a) to reduce the size of the diffraction pattern
ii. It is desirable	(b) to inspect an artery
iii. Special earplugs are needed	(c) to protect their hearing
iv. An **endoscope** is used	(d) to check your answer

> **endoscope** 内視鏡

Grammar point

Q2 Make full sentences describing a function from the phrases given.

i. Electrons – use – generate – - a magnetic field.

ii. A laser – continuously – monitor – the thickness of the product.

iii. Curved portions of highway sections – always banked (**tilted**) – prevent – cars from overturning.

> **tilted** 傾いている

iv. A heat pump – air conditioner – operated in reverse – heat a room.

Writing

Q3 Explain the different components of a computer operating system and describe their functions.

第 21 章　用途や機能の記述

　ものの用途あるいは機能は，対象物や研究プロジェクトの現実の世界でのはたらきを言及します。したがって，ある研究の果たす機能を詳細に論じる際，提案や批判，拒否，賛成，反対，問い，従来の研究についての議論，アドバイスなどを述べる必要があります。必ず，現在行っている研究の目的と有用性を述べることが求められます。

21.1　用途や機能の記述

　次の構文を用いて，機械の機能や装置の目的を説明します。

⑴ (in order) to＋不定詞

例：

・A voltmeter is used **to find** the potential difference between any two points in a circuit. （電圧計を用いれば回路中の任意の 2 点間の電位差がわかる。）

・**In order to monitor** solar flares, a space probe needs to be positioned along a line that extends directly toward the Sun. （太陽フレアを監視するには，太陽に向かって直接伸びる線に沿って宇宙探査機を配置する必要がある。）

⑵ for ＋ 動詞 ＋ –ing

例：

・Thomson's apparatus is used **for measuring** the ratio of mass to charge for an electron. （トムソンの装置は電子の質量電荷比を測定するために使用される。）

⑶ act as ＋ 名詞

例：

・Immediately after the switch is closed, the inductors **act as** broken wires. （スイッチを閉じると同時に，インダクタは断線として機能する。）

21.2　目的の表現

論文や口頭発表で研究活動の目的は，次のように述べられます。

⑴ aim/purpose/objective/goal:

例：

・We need to **aim** for precision in science. （科学では正確さが求められる。）

・One of the **goals** of physics is to document observations about our world. （物理学の目的の一つは，世界に存在するものについての観察を記録することである。）

・The primary **objective** is to obtain a nice correlation between its systems. （本研究の目的はシステム間の望ましい相間関係を得ることである。）

⑵ if/ whether と組みあわせる：

例：

・A related aim/purpose/objective/goal is to determine **whether** a force is affected by gravitation. （関連する目的・目標は，重力により力に変化が生じるかどうかを決定することである。）

⑶ that/which ＋ 現在形：

例：

・An example of an emf device is a thermopile **that provides** onboard electrical power in some spacecraft. （emf デバイスの例として，宇宙船に搭載電力を供給するサーモパイルがある。）

・The aim of the experiment, **which investigates** the effects of temperature, is to discern patterns and draw conclusions based on empirical data. （温度の影響を調べるこの実験の目的は，経験的なデータに基づいてパターンを見出し，結論を導くことである。）

テキスト：学会の重要性

　ネットワーキングとは，共通の職業や関心を持つ人々の間で情報やアイデアを交換することです。ネットワーキングは，専門家がその分野で起きている最新の出来事を知るのに役立つだけでなく，雇用の機会を見つける手助けをする機会にもなります。すぐれたネットワークはすぐれた人間関係の上に築かれます。良好なコミュニケーションは，すべての人間関係に不可欠な要素です。コミュニケーションは自分のニーズを満たすのに役立つだけでなく，前向きな気持ちや考え，示唆を仲間と共有することにつながります。したがって学会のような会議への出席は，研究者としての活動に欠かせません。学会への出席は，研究者が他の研究者と関わりを維持するための非常に重要な手段です。質の高い学会は，どのキャリアステージの科学者をもマンネリから脱却させる力をもっています。

　誰しも学会でのプレゼンテーションをすべて記憶することはできませんので，科学者は，他の人たちの知的好奇心を引きつけ，そのアドバイスから利益を得るためのあらゆる手段を探るべきです。専門的な学会は主に教授を対象としていますが，学生もまた，自分の研究テーマの進歩について学ぶだけでなく，自分のデータを投稿して発表することで，学会から利益を得ることができます。最も興味深く，重要なアイデアは常に書き留めておくべきです。

　他の人々との出会いは，共同研究の実現と創造性を可能にするに不可欠です。学会の主な利点は，フィードバックと同僚のサポートです。また，新しいトピックや出版物を知る上でも非常に役立ちます。さらに，多くの人にとって，人々と出会う楽しい機会でもあります。

Chapter **22**

Explaining strengths and weaknesses

Whenever presenting research, it is essential for the readers to realize that you are aware and **knowledgeable** of the strengths and weaknesses of your study. At the same time, you should have sufficient evidence to support your hypothesis. When presenting weaknesses of the study, the positive aspects should be highlighted while still being honest about the limitations and **flaws** present in the research.

knowledgeable よく知っている（= well-informed）

flaw 欠点

22.1 Strengths and weaknesses in research

An experimenter should ensure that their experimental design is **authentic** to make sure that the consequences and conclusions acquired are **valid**.

authentic 信頼できる，本物である
valid 有効である

22.1.1 Strengths of an experiment

An experiment is considered to possess strength if the following characteristics can be **guaranteed**:

guarantee 保証する（ギャラの語源）

- Ensuring that each of the variables that may have an effect on the results is controlled
- Use of a sample length which is **representative**
- Ensuring that the technique is written unambiguously so that if the experiment is repeated by someone else, there will be no confusion in understanding what to do

representative 代表的な

22.1.2 Weaknesses of an experiment

Usually a research process also suffers from some of the following <u>inherent</u> weaknesses that may not always be possible to completely **eliminate**.

inherent 固有の，内在する ⇒Ⓦ17
eliminate 取り除く
from time to time 時々
desirable 望ましい

- Experiments may be **from time to time** much less precise than **desirable**.
- Different types of effects in the laboratory are controlled as much as possible; however, this means that not all the effects that can affect outcomes are investigated during the research process.
- Experimenters usually have a few conclusions, at the beginning of the experimental process, about what will happen or what is supposed to happen and this may constitute a **bias** in the experiment with regard to <u>the reporting of</u> the results.

bias バイアス，先入観
<u>the reporting of</u> the -ing of の形をとる。例：The Making of the English Bible.

22.2 Explaining strengths and weaknesses of research in writing

No research, investigation or experiment is perfect. Therefore, not only the strengths of your work need to be expressed, but also the weaknesses and capacity for bias (e.g., in the methodology and data analysis). Particularly in the Discussion section, it is essential that you <u>purposively</u> provide the readers with opportunities for improvement considering the limitations of your own research, especially with regards to methodology and in the interpretation of the results. Such insights are a visible **cue** to your readers of the superior quality of your research.

<u>purposively</u> ⇒Ⓖ5.4

cue 手がかり

When writing about research results, you need to understand first the strengths and weaknesses of the evidence collected and present those simply. This is essential in showing that you **apprehend** the significance of what you have done and obtained. These results then need to

apprehend 理解している

be taken under consideration in developing your own argument about the answer to the research question. In this process, the reader has to be clear about your perspective. Possibly, your argument also has strengths and weaknesses as well - it is essential that you and the reader should understand these clearly. Readers prefer to be informed about any limitations or weaknesses in the research; on no account should the limitations of the research project be hidden. The problems may sometimes be even more interesting to the reader than the strengths.

on no account should the limitations of the research project be hidden
⇒ⓖ16.4

When analyzing data, a review of the research goals is always included since this helps to organize the data and focus the analysis. At this time, the strengths and weaknesses should be identified.

Language:

It appears to be or *it seems* – these phrases express the factual truth by stating the limitations of the research and yet presenting what is **intuitively** suspected or expected.

intuitively 直感的に

Examples:
· The light rays appear to come from a point behind the mirror.
· It seems that a single magnetic pole can be isolated.

💡 **Faults, flaws, and drawbacks:** A **fault** describes a weakness in something, such as mechanical, electrical or technical faults. **Flaw** is used to refer to a minor fault or weakness in something making it less effective. **Drawback** is used to refer to a feature of something which makes it less useful or acceptable than it could be. *Drawback* is often synonymous with disadvantage.

22.3   Explaining strengths and weaknesses of other's research

The important thing when discussing the strengths and weaknesses of other's work is not to **undermine** their credibility in any way. When highlighting the drawbacks of other's research, the following words are usually employed in a respectful manner:

undermine 損なう

inappropriate, complicated, **concern**, **conjectures**, *drawback, flawed, misleading,* **shortcomings**, **speculative**, *weakness*

concern 懸念（関心の意味でも用いる）
conjecture 推測
shortcoming 欠点
speculative 推測の段階,根拠を欠いている

Examples:
· Neither strategy works and thus is inappropriate to our purposes.
· The research goal has been difficult to achieve because light is complicated.
· There is a concern about the accuracy of the experimental results in this study.
· Of course, the authors' conjecture will be useful only if it **holds** when the applied force value is changed.

hold 成り立つ

Text: Laboratory experiments

The scientific method begins with a question about something you observe. In reality, the scientific method is a process of experimentation used to **investigate** observations. A hypothesis is the first step in the scientific method. It can be supported or refuted through carefully **crafted** experimentation. A hypothesis also includes an explanation of why the guess may be correct, and the experiments are performed to see whether the predictions are supported. The hypothesis and the resulting prediction that needs to be tested have to be clear before the experiment is started.

investigate 調べる
refute 反論する
craft 作成する

A science experiment is performed in a laboratory. Teachers have an **obligation** to instruct their students in the basic safety practices before using any apparatus. Maintaining complete and accurate records of all the details of a laboratory experiment is also very important. Before starting an experiment, the **independent variable** and **dependent variable** must be clearly established. Any variation in values must be small for data to be

obligation 義務

independent variable 独立変数
dependent variable 従属変数

considered reliable. Repeating a scientific investigation increases its reliability. Surprising results should be **replicate**d — both to make sure they are valid and to suppress people's doubts.

replicate 再現する

Scientific practice is characterized by its continual effort to test since science is knowledge gained through repeated experiment or observation. Multiple trials are necessary to see patterns in experimental data. Reflective observation is also essential from time to time. This means taking time-out from "doing" and stepping back from the task and reviewing what has been done and experienced.

interpretation 解釈
inference 推論
accuracy 精度
equipment 器具
apparatus [ӕpəˈrӕtəs] 装置

💡 **Keywords:** procedure, observation, **interpretation**, **inference**, hypothesis, experimentation, methods, results, **accuracy**, material, technique, investigation, prediction, construction, analysis, test, **equipment**, device, **apparatus**, set-up

Exercises

Understanding the text

Q1 Choose the word which best fits the meaning of the sentence.

i. _____ guide observation and experiment because they suggest what to observe. (*hypotheses* or *predictions*)

ii. Students _____ skills of practicing scientists in a laboratory. (*acquire* or *receive*)

Grammar point

Q2 Fill in the blanks with an appropriate word referring to weaknesses in another's study:

complicated, concern, misleading

i. The interpretation in this context is really _____.

ii. The exact form and location of the **charges**, as reported, are also of _____.

charge 電荷

iii. The conclusions reported do not hold for structures more _____, suggesting an inherent flaw in the hypothesis.

Writing

Q3 Write about the strengths and weaknesses of any product. Suggest some improvements.

第 22 章　強みと弱点の説明

　研究発表を行う際，論文の著者はその研究の強みと弱点を自覚していることを読者に知らせなければなりません。同時に，立てた仮説を支持する十分な証拠はもっていなければなりません。研究の弱点を示す際には，肯定的な一面を強調する一方で，研究のもつ限界や欠点についても率直に述べるべきです。

22.1　研究の強みと弱点

　実験では全工程に誤りがないことを示さなければなりません。でないと，結果と結論に信憑性がなくなってしまいます。

22.1.1　実験の強み

以下のような要件が満たされている場合，すぐれた実験とみなされます。

・実験の結果を左右する変数がすべて事前に計算されたものであること（＝偶然生じたものではないこと）

・代表的なサンプル長さの使用
・実験手順がすべて明確に述べられており，他の研究者が同じ実験を行っても何ら支障をきたさないこと

22.1.2　実験の弱点

多くの場合，実験の工程には以下のような内在的な不具合な面があり，それを取り除くのは容易ではありません。

・実験の精度がかなり劣ってしまい，望ましくない，ということが時に起きてしまう
・実験の結果に影響を与えうる要素は可能な限り制御される。ただし，これは，結果に影響を与える可能性のあるすべての要素が調査されているわけではないことを意味する
・実験を開始する時点で，実験者は通常，実験で何が起こるか，または起こりうるかについていくつかの見通しをもっている。これは，実験結果の報告の際にバイアス[*1]をもち込む可能性がある

*1 自分に都合のいいように議論を展開すること

22.2　行った研究の強みと弱点を文章で説明する

いかなる研究，調査，実験も完全ということはありませんので，行った研究のすぐれた点だけでなく，限界にも触れるべきです。限界の中に，実験方法やデータの分析でのバイアスがあります。特に考察セクションで今回の研究の方法や結果の解釈などのもつ限界に触れ，読者にその点の改善を呼びかけることが重要です。そうすることで，読者はあなたの研究の質を高く評価するでしょう。

研究結果を記す際，得られたデータの強みと弱点をまず認識し，簡潔に述べる必要があります。実験結果の意義を理解していることを読者は知ることになりますので，強みだけでなく弱点にも言及することは非常に重要です。次に，実験結果を研究課題への解答との関連で述べます。このプロセスは，読者に明確に伝わるように述べます。その際，議論の展開にも強みとともに弱点が顔を出すことがありえます。このことを著者も読者もはっきり認識する必要があります。読者は研究の限界や弱点について正直に述べてほしいと思っていますので，弱点を隠すことがあってはなりません。強みよりも問題点の方が読者にとっては興味深いこともあります。

データを分析する際，研究の目的にも常に言及してください。それにより，データがうまくまとめられ，分析が明確に行われます。ここでも，強みと弱点に触れてください。

言語表現

It appears to be / it seems：研究の限界を述べ，しかし直観的には，正しいのではと思っているデータを提示する際に用いられます。

例：
・**The light rays appear to** come from a point behind the mirror.（光線は鏡の後ろの点から来ているように見える。）
・**It seems** that a single magnetic pole can be isolated.（単一の磁極を分離できるようだ。）

> **faults, flaws, drawbacks:** fault は機械や電気，技術などの弱点を指します。flaw は，効果を薄める小さな欠点を指します。drawback は disadvantage 同様，製品などが，予想されたほどの有用性や満足度を達成していない場合に用いられます。

22.3　他の研究の強みと弱点の説明

他の研究の強みと弱点を論じる際に，その研究の価値を損なうようなことは決してしてはなりません。他の人の研究の弱点を強調する場合，次のようなことばを用い，失礼にならないよう述べてください。

complicated, concern, conjectures, drawback, flawed, inappropriate, misleading, shortcomings, speculative, weakness

例：
・Neither strategy works and thus is **inappropriate** to our purposes.（どちらの戦略も機能しないため，本研究の目的には不向きである。）
・The research goal has been difficult to achieve because light is **complicated**.（光が複雑なため，研究目標の達成は困難だった。）
・There is a **concern** about the accuracy of the experimental results in this study.（この研究の実験結果には正確さについて懸念がある。）
・Of course, the author's **conjecture** will be useful only if it holds when the applied force value is

changed.（もちろん，著者の推測が有効なのは，加えられた力の値が変更されたときにそれが成り立つ場合のみである。）

テキスト：実験

　科学的方法は，観察する対象について問いを発するところから始まります。実際には，科学的方法とは一連の実験を通して観察結果を考察することをいいます。

　科学的方法の第一歩は仮説を立てることです。周到に準備された実験によって仮説を支持，あるいは否定することができます。仮説にはまた，立てた予測が妥当な可能性があることの説明が含まれています。そして実験を行って，その予測が正しいかどうかを検証します。仮説と仮説から導かれる予測を検証可能なものにするには，実験を行う前にそれらを明確な形にしておかねばなりません。

　科学実験は実験室で行います。実験用具の安全な使用について，教師は用具を使う前に学生にきちんと指導しなければなりません。実験の詳細はすべてもらさず正確に残しておくこともとても重要です。開始前に，独立変数と従属変数を確定しておかなければなりません。実験結果の値のばらつきが少ないものが，信頼できるデータです。科学調査を多く繰り返せば，それだけ信頼性が高くなります。予想しなかった結果に終わった実験は，実験への信頼性を高め，他の研究者の疑念を払うため，再現できるものでなければなりません。

　科学は繰り返し行われる実験や観察により得られる知なので，科学的実践は継続的なテストへの取り組みのうえに成り立ちます。実験データから何らかの法則性を導き出すには繰り返しの試行が必要です。適当な間隔で，行っている観測を振り返ることも必要です。振り返りとは，一旦作業を停止し，これまでの実験を見直すことです。

> **キーワード：** procedure, observation, interpretation, inference, hypothesis（仮説）, experimentation（実験）, methods（実験の手法）, results, accuracy, material, technique investigation, prediction（予測）, construction, analysis, test, equipment, device, apparatus, set-up

Chapter 23

Agreeing and disagreeing

When writing a research paper or giving a presentation, you will **eventually be compelled to** agree or disagree with something already published. However, just because you disagree with someone does not mean that you should **rule out** the possibility of a friendly and professional relationship with that person in the future. So, it is essential to use the proper language which **precludes** any type of impoliteness. This chapter discusses the language we use in professional English when we want to agree or disagree with someone.

eventually 最終的に
be compelled to … ～しなければならない（～を強制されるが基本的な意味）
rule out 排除する
preclude 排除する ⇒Ⓦ6
（pre-（前もって）+ clude
= close（閉じる））

23.1 Agreement and disagreement

While expressing your own views, opinions, or suggestions in the Discussion section of a research paper, it is **inevitable** that your conclusions will not be in agreement with other researchers' results and interpretations. Some may have the same views, while some may draw different **inferences**. Therefore, when writing a research paper, you may be required to accept or reject the suggestions and assumptions of others. Agreement should be expressed using words like *exactly* and in an energetic tone, whereas disagreement should always be expressed politely. If you disagree with what a previous researcher has said, you should feel free to do so, but do so in a modest and **amicable** manner. The ability to persuade others is an important skill which must be developed through the use of valid arguments in support of your position. In writing, the major issues should be outlined first and then your viewpoint should be stated. That is, the justification should be expressed first and the conclusion later. If your opinion is stated first before the reasons, the evaluator will only get an impression of what you think, not the process by which you arrived at your conclusion.

inevitable 避けられない

inference 推論

amicable 親しい，友好的な

by which ⇒Ⓖ8.1

💡 Words expressing disagreement with other's results: *contrast, disagreement, **discrepancy**, in-compatibility*, *variation*, <u>*discordance*</u>

discrepancy 相違
incompatibility 両立しがたいこと
discordance 不一致⇒cord
と heart の関係（cord =
heart）はグリムの法則（Ⓦ
1）の一例

23.2 Agreeing in conversation

It is always a good idea to justify your points of view. Do not simply say "I agree", but rather "I agree because I think that ... (explain your reasoning)." <u>Below are some phrases</u> used when agreeing with someone; it is always better to be as enthusiastic as possible when in agreement.

below are some phrases
⇒Ⓖ16.4

Examples:
· Absolutely/Definitely/Exactly
· I can't help thinking the same.
· **I couldn't agree more.**
· I feel the same.
· I (totally) agree with you/that.
· No doubt about it.
· That's a good point/ I see your point.
· That's just what I was thinking.
· That's my view exactly.
· That's right.
· I see where you're coming from.
· True enough.

I couldn't agree more. 大賛成です（これ以上賛成しようとしてもできない）

Table 23.1

Phrases of agreement	Degree of agreement
· Absolutely/definitely/of course · I also think so. · I agree with you a hundred percent. · I agree with you entirely.	complete agreement
I have no objections.	People usually say this when they are <u>not</u> completely committed to <u>some</u>thing but do not see why they should oppose it either.

<u>not</u> ... <u>some</u> ⇒Ⓖ3.5

23.3 Disagreeing in conversation

Any professional relationship includes both agreements and disagreements. What matters, particularly when disagreeing with someone, is how you disagree. Remember to keep emotional expressions to a minimum during an argument or disagreement. Agreements and disagreements are usually about your personal thoughts and feelings about something. Phrases such as *I think* or *in my opinion* make it clear that you are expressing an opinion rather than a fact. Using phrases like these helps to **foster** a welcoming environment in which people feel respected and free to express themselves. *I say this with due respect, but* ... is a **courteous** way of disagreeing, especially in a professional or formal setting.

foster 育てる

courteous [ˈkɜːrtiəs] 丁重な

The following are some phrases and expressions to use when disagreeing with others:

- · But what about ...,
- · Don't you think it would be better ...,
- · Frankly, I doubt if ...,
- · I am afraid I do not agree ...,
- · I do not agree, I would prefer ...,
- · I don't think that ...,
- · I'm afraid I don't agree ...,
- · Let us face it, the truth of the matter is ...,
- · Shouldn't we consider ...,
- · The problem <u>with</u> your point of view is that ...

<u>with</u> ⇒Ⓖ7.5

Table 23.2

Words and phrases	Degree of disagreement
· I beg to differ. · I think there's a better explanation.	A strong, formal and polite phrase used to express disagreement
· Yes, but ...	Some points are partially agreed upon, but may not be entirely agreed upon.
· To be honest, · I don't agree with that.	A polite way to express complete disagreement.
· As a matter of fact · I don't think that's correct.	A firmer but more formal way of expressing disagreement.

Text: Environment

All energy sources have some impact on our environment. Ships and airplanes, **large industrial operations**, **coal-fired power plants**, and cars and trucks are major sources of nitrogen oxide emissions. As a result, tiny aerosol particles can be found all over the world, including the oceans, deserts, mountains, and forests. Despite their small size,

large industrial operation 大規模な企業活動
coal-fired power plant 石炭火力発電所

they have a significant impact on our climate and our health as they move through the atmosphere. Wildfire smoke is also a very complex type of air pollution. Rain droplets absorb air pollution elements like **sulfur** and **nitrogen** and winds then spread these acidic **compound**s through the atmosphere and over hundreds of kilometers. Waste was dumped in rivers and in the sea before factories were forced to stop damaging the environment.

Engineers must therefore produce and **implement** innovative solutions at a reasonable cost while maintaining environmental sensitivity. The advantages of geothermal power plants, <u>for example</u>, are the same as those of nuclear power plants. Everyone has to be involved in efforts to save the environment by controlling the **pollutants** of air and water since climate change and environmental <u>degradation</u> undermine the rights of every human. Instead of non-renewable sources which will be **depleted** in the future, renewable sources of energy like solar and **tidal** energy, the power of the wind, **subterranean** hot waters, and so on, need to be utilized. Undoubtedly, **stricter** environmental **regulations** have <u>compelled</u> many businesses to use cleaner energy sources.

> Different words and phrases can be used to highlight a contrast between two ideas, such as: *although*, *even though*, *despite*, *in spite of*, *however* and *but*.

sulfur 硫黄
nitrogen 窒素
compound 化合物

implement 実行にうつす
for example ⇒ G 16.3
pollutant pollution（汚染）を引き起こすもの。(-ant = -ing)
degradation 悪化 ⇒ W 8
deplete 使い果たす（de- (down) + plete（= plenty, full））⇒ W 8
tidal 波の
subterranean 地下の
stricter より厳しい
regulation 規制
compel ⇒ W 7

Exercises

Understanding the text
Q1 What are the positive and negative aspects of using nuclear energy sources?

Grammar Point
Q2 Write whether the following phrases mean (a) I agree or (b) I disagree?

  i. OK, I can go along with that.

  ii. That seems about right.

  iii. I would differ with you there.

  iv. You're absolutely correct.

  v. You have a good point there.

  vi. I am in complete agreement with you.

  vii. I am not sure about that deadline.

Writing
Q3 Write about an environmental problem and discuss one solution being proposed by scientists. Then agree or disagree with this proposal, giving your reasons clearly.

Chapter 23

第 23 章　賛成と反対意見を述べる

　論文執筆あるいは発表する際，最終的に先行研究に同意するか反対するかのどちらかの選択を迫られます。しかし，反対することが研究者とのこの先の友好な人間関係を妨げるものであってはなりません。したがって，反対する際に礼を欠いた表現を慎むよう十分気をつけてください。本章では，専門分野で使う英語において相手の意見を支持するときや支持しないときに使う言葉について説明します。

23.1　賛成と反対

　学術論文の考察セクションで自分の見解や意見，示唆を述べる際に，あなたの導いた結論が他の研究者の結論や解釈と食い違うことは必ずあります。同じ見解の部分もあれば，推論が異なる部分もありえます。したがって，学術論文執筆の際，他の研究で示された示唆や仮定への賛成や反対を述べる必要があるかもしれません。同意する場合は，exactly といった表現を選び力強い口調で述べます。反論する際には他の研究への敬意を失わないことが大切です。先行研究に反論する際は遠慮なくすべきです。しかし，おだやかな，相手への敬愛の念をもってなされるべきです。人を納得させる議論の展開の仕方はぜひ身につけたい技術です。それは，自説を展開するときの他の研究への敬意を失わない論の展開などにより培われます。

　執筆の際，研究の主要課題をまず述べ，その後にあなたの見解を示してください。取り上げる課題の重要性をまず示し，そしてそれに対する結論を述べるのです。意見を理由の先に提示するという形をとると，評価する人は，あなたがどう考えているかという印象を受け取るだけであり，あなたがその結論に至った過程を知ることはできないのです。

> 💡 他の研究の結果に賛成できない時に用いられることば：contrast, disagreement, discrepancy, incompatibility, variation, discordance

23.2　口頭発表の際の賛成の仕方

　自分の意見の正当性を述べるのはよいアイデアです。ただ「賛成です」と発言するのではなく，「こうこういう理由で賛成します」と言いましょう。以下，賛成の表現を紹介します。賛成する際には，できる限りつねに熱意をもって臨んでください。

例：
- absolutely/definitely/exactly（まったくその通りです。）
- I can't help thinking the same.（そのように考える他ありません。）
- I couldn't agree more.（大賛成です。）
- I feel the same.
- I (totally) agree with you/that.
- No doubt about it.
- That's a good point/ I see your point.（おっしゃる通りです。）
- That's just what I was thinking.
- That's my view exactly.
- That's right.
- I see where you're coming from.
- True enough.

表 23.1

賛成の表現	賛成の程度
・absolutely/definitely/of course ・I also think so. ・I agree with you a hundred percent. ・I agree with you entirely.	100 ％の賛成
I have no objections（反対）.	積極的に賛成とはいえないが，反対する理由もないときに言う。

23.3　学会や会話の場での反対意見の述べ方

　あらゆる仕事上の人間関係において，賛成反対の両方が出てきます。反対の場合には，その意見の述べ方が問題となってきます。極力感情的なことば遣いにならないよう十分に気をつけてください。賛成，反対という判断は，人の考えや感情と密接に結びついています。"I think" や "in my opinion" といった表現は，事実ではなく意見を述べていることを示します。この前置きがあれば，敬意の念が相手に伝わり，相手は自分の意見を口にすることができます。"I say this with due respect, but …"（謹んで申し上げます。しかし，…）という表現は，相手への配慮の行き届いた言い方で，仕事上の，あるいはあらたまった場では特に重要です。

　以下に示すのは，他人と意見を異にするときに使うフレーズや表現です。

- But what about …,（こういうのはいかがでしょうか。）

・Don't you think it would be better ...,
・Frankly, I doubt if ..., （率直に申し上げますと，～とはならないのではないでしょうか。）
・I am afraid I do not agree ...,
・I do not agree, I would prefer ...,
・I don't think that ...,
・I'm afraid I don't agree ...,
・Let us face it, the truth of the matter is ..., （率直に申し上げます。）
・Shouldn't we consider ...,
・The problem with your point of view is that ...

表 23.2

表現	反対の程度
・I beg to differ. （失礼ですが，私の考えは違います。） ・I think there's a better explanation.	意見の違いを述べる際に用いられる，強くてあらたまった，丁寧な表現
・Yes, but ...	あるところまでは賛成だが，100% 賛成とはいえない
・To be honest, ・I don't agree with that.	おだやかに反対の意を述べる
・As a matter of fact ・I don't think that's correct.	より強い，あらたまった反対の意見

テキスト：環境

　すべてのエネルギー資源は，環境に影響を与えます。窒素酸化物を主に排出するのは船舶や飛行機，大規模な企業活動，石炭火力発電所，車やトラックです。その結果，微小のエアロゾル粒子が海や砂漠，山，森を含む世界のいたるところで観察されています。微小な物質ではありますが，大気を移動し，気候や人の健康に影響を与えます。山火事の煙も大気汚染の一因となります。雨の水滴が硫黄や窒素などの大気汚染物質を吸収し，風がこれらの酸性化合物を何百キロメートルにもわたり大気中に拡散させます。工場が環境破壊の停止を余儀なくさせられるまでは，廃棄物は川や海に捨てられていました。

　したがって技術者は，コストを考慮に入れ，かつ環境への影響を配慮したうえで，革新的な解決策を生み出し実行しなければなりません。たとえば地熱発電は，原子力発電と同じような効率で電力を生み出します。私たちは皆，空気や水を汚染する物質を極力出さないことで環境の維持に努めなければなりません。気候変動と環境の劣化は人間の暮らしの質を大幅に下げてしまいます。いずれ枯渇しかねない再生不可能なエネルギーに頼るのではなく，太陽エネルギーや波力エネルギー，風力エネルギー，地熱エネルギーなどの再生可能なエネルギーを利用する必要があります。より強固な環境保護政策を通して，企業はよりクリーンなエネルギー源を使うようになったことは間違いありません。

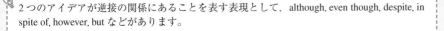

　2つのアイデアが逆接の関係にあることを表す表現として，although, even though, despite, in spite of, however, but などがあります。

Chapter 24

Cause and effect

A cause is something that produces an event or condition; an effect is something that happens as a result of an event or condition. In a research paper, it is an important goal to determine how various **phenomena** relate in terms of origins and outcomes.

; ⇒ⓖ16.2.1

phenomena 現象（phenome-non の複数形）

24.1 Expressions showing cause and effect

Events and situations are linked together by causes and effects. Causes explain why something occurs, whereas effects describe the results.

(1) Clauses

Dependent clauses, which begin with subordinating **conjunctions** like "while," " that," or "unless," provide context but cannot stand on their own. Subordinating conjunctions thus link an independent clause to a dependent clause, showing cause and effect.

dependent clause 従属節
(= subordinate clause)
conjunction 接続詞

Examples:
- Because it deals with how the universe works, physics is the most fascinating subject in the world.
- Since acceleration is a **constant**, it can be calculated outside of the **integration**.
- As I was exhausted, I went to bed early.

constant 定数
integration 積分

(2) Phrases:

An adverbial phrase is similar to an adverb in that it adds more information to the sentence, but it describes the verb with more than one word.

adverbial phrase 副詞句
in that ⇒ⓖ7.1

Examples:
- The boat came to a stop as a result of a wind shift.
- Einstein demonstrated in 1905 that, as a consequence of his theory of special relativity, mass can be thought of as another form of energy.
- We calculated the Doppler shift in wavelength due to the motion.
- On account of experimental errors in weighing and measuring, such analyses are rarely **trustworthy**.
- Trade is not large, however, owing to the costs of transportation.

trustworthy 信頼できる
however ⇒ⓖ16.3

(3) Connectors

A cause in one sentence or clause can be linked to an effect in the following sentence or clause by a connector.

connector コネクタ（接続語）

Examples:
- I think, therefore I exist.
- Accordingly, this energy is rapidly **dissipated**.
- As a consequence, mass can be thought of as another type of energy.
- Hence, the rotational inertia of the rod is much smaller.
- Because of this property, entropy change is sometimes referred to as "the **arrow** of time."
- Consequently, the **magnetic field** is nearly **uniform**.
- As a result, the wave is described as **plane-polarized**.
- I know you must be tired, so I will let you rest.
- That is why the time interval measured between the two events was longer.
- That is the reason why you measured a length that was less than the correct length.

be dissipated 散逸する

hence ⇒セクション 6.4.1
arrow 矢

magnetic field 磁場
uniform 均一な
plane-polarized 平面偏光
⇒ⓖ16.1

- According to special theory of relativity, motion affects the measurement and **thus** reality.

💡 **Words describing cause and effect:** *account, because, consequence, consequently, due, reason, result, so, why*

---

(4) Verbs and verb phrases

Sentence structure: Cause + verb + effect

Examples:
- The greater speed led to a loss of control in the **landing**.
- A mistake may result in serious **complications**.
- The gravitational potential energy can be used to account for the jumper's motion.
- Any sudden air pressure reduction can bring about nitrogen-forming bubbles in the blood.
- According to thermodynamics, this **backward process** gives rise to a decrease in entropy.
- The interference will be responsible for the brightest possible illumination.

landing 着陸
complication 事態

backward process 逆の操作（進行）

A be responsible for B
A が B をもたらす
⇒ © 16.9.3

Sentence structure: Effect + verb + cause

Examples:
- The change in **kinetic energy** is the result of a change in speed.
- They arise from the relativistic mass-energy equivalence formula.
- The pressure on the container wall is attributable to oxygen.
- Dark regions stem from fully destructive interference.

kinetic energy 運動エネルギー

attributable to ...
[əˈtrɪbjətəbl] ～に起因する

24.2   Degrees of possibility of the cause

In a scientific investigation, researchers may have to evaluate the degrees of possibility of different potential causes of a phenomenon by considering factors such as experimental results, theoretical models, and prior knowledge.

	Possibility (75%)	Certainty (100%)
Present	Example: ・ The line may be vertical, horizontal, or **slanted**.	Example: ・ The line must be perfectly straight.
Past	Example: ・ The functions may have had to be interchanged.	Example: ・ The particle must have had an acceleration.

slanted 斜め

24.3   Uncertain causal connection

Sometimes there is a **correlation** between two events, that is, one event always accompanies the other, although there may or may not be a **causal connection**.

correlation 相関
causal connection 因果関係

Example:
- If/when/as all wavelengths are mixed, white light is produced.

Text: Energy sources

Every person, animal and device uses energy. There are numerous energy sources. Chemical energy, electrical energy, **electromagnetic radiation** (light), heat/ thermal energy, mechanical energy, and nuclear energy are all forms of energy. As a result, the energy we use to power everything from our homes to schools and workplaces is derived from a variety of sources. These can be divided into two categories: renewable and non-renewable energy sources. Primary energy sources include fossil energy such as oil, coal and natural gas, nuclear energy, and renewable energy sources such as geothermal, hydropower, wind, and solar. Electricity is a secondary energy source produced from primary energy sources. In the U.S., 20% of electricity is produced from nuclear energy, and about 20% from renewable energy sources. Although its share of total energy consumption has **declined**, oil remains Japan's most important source of primary energy.

On a worldwide scale, fossil fuels (coal, natural gas, and oil) currently provide the great majority of our energy, accounting for roughly 81% of what we use. The supply of energy sources which are nonrenewable is constrained by what we can **mine** or remove from the earth. Over thousands of years, coal, natural gas, and **petroleum** were produced from the **buried** remains of ancient marine plants and animals that lived millions of years ago.

Innovation has enabled societies to access a wide range of energy sources. Renewable energy comes from naturally renewing sources, making renewable resources nearly endless. These energy sources are **abundant**, sustainable, and environmentally friendly.

> **Relevant vocabulary about energy:** abundant, available, collect, consider, constitute, contribute, **convert**, **crude**, expensive, **extract**, generate, release, significant, **unrefined**, **viable**

electromagnetic radiation 電磁放射線

is derived from … ～に由来する，～から生ずる

consumption ⇒ Ⓦ16
decline 減少する
accounting for ⇒ Ⓖ16.9.1
constrain 制約を設ける
mine 採掘する
petroleum 石油
buried ['berid] 埋められた
abundant 豊富にある
convert 変換する
crude 加工していない
extract 抽出（する），抜粋（動詞では a を，名詞では e を強く読む）
unrefined 精製されていない
viable 実現可能な

Exercises

Understanding the text

Q1 Fill in the following blanks.

energy, fossil, global, plentiful

(a) _____ fuels　　(b) _____ sources

(c) _____ supply　　(d) _____ warming

Grammar point

Q2 Match one part of a sentence from A and <u>one</u> from B to form sentences of cause and effect.

one ⇒ Ⓖ3.2

A	B
i. Color can only be **duplicated** from one angle by copy machines	(a) as a result of the symmetry.
ii. Forces acting in all other directions will **cancel** one another	(b) and, as a result, any color shift caused by a change in perspective cannot be duplicated.
iii. The Doppler effect increases the frequency	(c) because of the car's motion.

duplicate 複製する

cancel 相殺する

Q3 Choose the correct phrase in each of the following.

i. The change in internal energy of the gas is the same for all three processes that *result from/result in* the same temperature change.

ii. The entropy of the substance changes *as a result of/giving rise to* the energy transferred to it.

iii. There is no **induced charge** on the spheres *resulted in/because of* their large separation.

induced charge 誘導電荷

iv. The **electric field** is *caused by/results in* a line of charge.

electric field 電場

v. The force *accounting for/due to* any ring element will be canceled.

Writing

Q4 Write about how two energy sources can contribute to decreasing the consequences of global warming.

第 24 章　原因と結果

　原因は，ある事象や状態を作り出すものです。結果は，ある事象や状態により生じるものです。学術論文では，さまざまな現象がどのように関係しあって最初の状態から最終の状態に移行したかを突き止めることが大切です。

24.1　原因と結果を表す表現

　原因と結果は事象と状況を結びつけます。原因はなぜ起きたかを述べ，結果は起きた結果の状況を表します。

(1)　**節**

while や that，unless などの従属接続詞で始まる従属節は文脈を提供しますが，独立した文を構成することはできません。このように従属接続詞は，主節と従属節を結びつけ，原因と結果を示します。

　例：
- **Because** it deals with how the world works, physics is the most fascinating subject in the world. (物理学はこの世でもっとも興味深い学問だ。世界のはたらきを解明する学問だからである。)
- **Since** acceleration is a constant, it can be calculated outside of the integration. (加速度は一定の値（定数）なので，積分の式から外して扱うことができる。)
- **As** I was exhausted, I went to bed early. (くたくただったので，早く寝た。)

(2)　**句**

文の情報を豊かにするという点で，副詞句は副詞に似ています。しかし，副詞句は 2 語以上の語句で動詞を修飾します。

　例：
- The boat came to a stop **as a result of** a wind shift. (風向きが変わり，舟は静止した。)
- Einstein demonstrated in 1905 that, **as a consequence of** his theory of special relativity, mass can be thought of as another form of energy. (質量がエネルギーのとる一つの形態でありうることを，アインシュタインは 1905 年に特殊相対性理論の帰結の一つとして明らかにした。)
- We calculated the Doppler shift in wavelength **due to** the motion. (動きによる波長のドップラー効果を計算した。)
- **On account of** experimental errors in weighing and measuring, such analyses are rarely trustworthy. (計量と測定に実験誤差があり，このような分析が信頼できることはほとんどない。)
- Trade is not large, however, **owing to** the costs of transportation. (しかし輸送にかかる費用が足枷となり，取引の量には限界がある。)

(3)　**コネクタ**

文と文，あるいは節と節の間の因果関係はコネクタ（接続語）によって表されます。

　例：
- I think, **therefore** I exist. (我思う，ゆえに我あり。)
- **Accordingly,** this energy is rapidly dissipated. (それに伴い，このエネルギーは急速に散逸される。)
- **As a consequence**, mass can be thought of as another type of energy. (その結果，質量をエネルギーが別の形で存在したものだとみなすことができる。)
- **Hence**, the rotational inertia of the rod is much smaller. (その結果，棒の回転慣性はずっと小さい。)
- **Because of** this property, entropy change is sometimes referred to as "the arrow of time." (この性質の

ため，エントロピーの変化は「時間の矢」と呼ばれることがある。)
- **Consequently**, the magnetic field is nearly uniform. (したがって，磁場はほぼ均一である。)
- **As a result**, the wave is described as plane-polarize. (したがって，波は平面偏光といわれる。)
- I know you must be tired, **so** I will let you rest. (疲れているにちがいなから休ませてあげよう。)
- **That is why** the time interval measured between the two events was longer. (だから，その 2 つの事象の間の時間差はもっと大きくなった。)
- **That is the reason** why you measured a length that was less than the correct length. (そういう理由で君の計測は誤って短くなってしまった。)
- According to special theory of relativity, motion affects the measurement and **thus** reality. (特殊相対性理論によれば，運動は測定に影響を与え，その結果現実にも影響する。)

> 💡 原因と結果を表す言葉：account（説明），because（なぜなら），consequence（結果），consequently（その結果），due（原因），reason（理由），result（結果），so（だから），why（なぜ）

⑷　動詞と動詞句

文の構造：原因＋動詞＋結果

例：
- The greater speed **led to** a loss of control in the landing. (落下時のスピードが速くなり，着地の際のコントロールがきかなくなった。)
- A mistake may **result in** serious complications. (一個のミスが重大な結果を引き起こしかねない。)
- The gravitational potential energy can be used to **account for** the jumper's motion. (重力位置エネルギーを考慮に入れることで跳躍者の動きを説明できる。)
- Any sudden air pressure reduction can **bring about** nitrogen-forming bubbles in the blood. (空気圧が突然わずかでも減少すると，血中に窒素を作る泡を生み出してしまう。)
- According to thermodynamics, this backward process **gives rise to** a decrease in entropy. (熱力学によれば，この逆過程はエントロピーを減少させる。)
- The interference will be responsible for the brightest possible illumination. (この干渉が最大限の照度をもたらす。)

文の構造：結果＋動詞＋原因

例：
- The change in kinetic energy is the **result of** a change in speed. (運動エネルギーの変化はスピードの変化により生じる。)
- They **arise from** the relativistic mass-energy equivalence formula. (それらは質量とエネルギーの関係を示す相対性理論の等価式により生じる。)
- The pressure on the container wall is **attributable to** oxygen. (容器壁にかかる圧力は酸素に起因する。)
- Dark regions **stem from** fully destructive interference. (暗い領域は完全な破壊的干渉により生じる。)

24.2　原因の強弱

　科学的調査において，研究者は，実験結果，理論モデル，予備知識などの要因を考慮することによって，現象のさまざまな潜在的原因の可能性の程度を評価しなければならない場合があります。

	可能性（75%）	確実性（100%）
現在形	例： ・The line **may** be vertical, horizontal, or slanted. (線は垂直，水平あるいは斜めのいずれかで存在しうる。)	例： ・The line **must** be straight. (その線はまっすぐであるにちがいない。)
過去形	例： ・The functions **may** have had to be interchanged. (これらの機能は入れ替えが必要だったかもしれない。)	例： ・The particle **must** have had an acceleration. (粒子は加速度をもっていたにちがいない。)

24.3　因果関係がはっきりしない場合

2つの事象の間に因果関係があるのかはっきりしないが，それらに相関関係がある，つまり，ある事象の後に必ず別の事象が付随する，ということがあります。

例：

・If/when/as all wavelengths are mixed, white light is produced.（すべての波長が入り混じると白い光が発生する。）

テキスト：エネルギー源

　人間や動物，機械はエネルギーを消費します。エネルギー源は数多く存在します。化学エネルギー，電気エネルギー，電磁放射線（光），熱エネルギー，力学的エネルギー，原子力エネルギーはすべて，エネルギーの形態です。その結果，家庭や学校，職場でに至るまで，あらゆるものに電力を供給するために使用されるエネルギーは，さまざまな供給源から得られています。それらは再生可能と再生不可能なエネルギー源に大別されます。主要なエネルギー源は，化石エネルギー（石油，石炭，天然ガス），原子力エネルギー，再生可能エネルギー（地熱，水力，風力，太陽光）です。電力はこれら主要エネルギー源から発生する二次的エネルギー源です。アメリカ合衆国では，電力の20%は原子力エネルギーから，20%は再生可能エネルギーから生み出されています。日本では，消費量は減ってきているとはいえ，石油が最も重要なエネルギー源となっています。

　地球全体では，現在，化石燃料（石炭，天然ガス，石油）が私たちが使うエネルギーの大半（約81%）を生み出しています。これらのエネルギー源は地中から取り出す量に限度があるため，再生不可能なエネルギーとされています。石炭，天然ガス，石油は，何百万年も前に生存した海洋の動植物から何千年もかけて作られました。

　技術革新によりさまざまなエネルギー源の使用が可能になっています。再生可能エネルギーは自然に再生されるものから生じ，事実上無尽蔵ということになります。これらのエネルギー源は豊富であり，持続可能であり，環境を傷つけません。

> **エネルギーを論じる際に使われる語句**：abundant（豊富な），available, collect, consider, constitute, contribute, convert, crude, expensive, extract, generate, release, significant, unrefined, viable

Chapter 25

Certainty and possibility

When writing about research, we sometimes have to express <u>a strong possibility</u> that something is true, <u>especially</u> when we lack actual evidence. Sometimes, we speak about things that are possible in general. Also, we may be unsure about some aspects of our study but believe it should be possible. Finally, sometimes we can express complete certainty. So, all of these different situations require the use of modal verbs with slightly different meanings, and these are described in this chapter.

a strong possibility ⇒Ⓖ2.5.1
especially ⇒Ⓖ5.5

25.1 Scale of certainty

We can **differentiate** between the following categories by considering that the scale of cetainty ranges from 100 % to 0 % (The values are not exact).

certainty (100 %), *probability* (75 %), *possibility* (50 %), *improbability* (10 %), *impossibility* (0 %)

differentiate 違いを設ける

Table 25.1 Language for each category

Category	Examples
Certainty (100 %)	· We are absolutely certain/positive/sure that torque and **angular momentum** are related. · The temperature is unquestionably higher than the Curie point. · The forces cannot cancel because one is certainly/definitely greater.
Probability (75 %)	· The powder explosion will most/very/quite likely occur within the pipe. · This is the most probable distance between the **single electron** and the central **proton**. · In a scientific experiment, increasing the concentration of the reactant could lead to a higher reaction rate..
Possibility (50 %)	· The distance you **perceive** may/might not be the same as the true distance.
Improbability (10 %)	· Momentum should not be confused with energy. · Slope measurement will probably not be precise. · It is unlikely/improbable that you will see upside-down racing.
Impossibility (0 %)	· A **flick of the finger against the wall** will certainly/definitely not cause damage. · The **intrinsic** properties of the system cannot possibly change. · In **differentiating** a **vector product**, be sure/certain/positive not to change the order of the two quantities.

angular momentum 角運動量

single electron 陽子
proton 電子

perceive 知覚する

flick of the finger against the wall 壁を指ではじく
intrinsic 固有の
differentiate 微分する
vector product ベクトル積

25.2 Conditional sentences expressing possibility

The conjunction *if* is used in two ways:

(1) The event is <u>a real possibility</u>:

Example:
· If an object is **subjected to** a force, then it will experience acceleration.

conditional sentence 条件文

a real possibility ⇒Ⓖ2.5.1

subjected A to B A を B の影響下に置く

(2) The event is an impossibility:

Example:
· If the lines were not **parallel**, an element arriving at their intersection would have two different **velocities** at the same time.

parallel 平行な
velocity 速度

25.3 Words and phrases expressing possibility

A discussion must give the reader the opportunity to debate or discuss what has been stated. Even if it is a good argument, a paper or report which **expresses** too much **assurance** is not good. There should be a range of certainty. The following are some examples of how to represent assumption and assurance:

expresses … assurance 自分の論は間違いないと断定する

(1) These expressions are used at the beginning or at the end of a sentence:

We *think/believe/guess/suppose/assume*; *In our opinion*; We are *certain/sure*

(2) These linguistic patterns can be used to make educated guesses about the past when you do not have all the relevant facts.:

educated guess 知識や経験に基づく，信頼度の高い推測

(A) To imply potential, use *may/ might/ could* + have + perfect infinitive.

Example:
· Some of you may have already read that chapter.

(B) To express assurance, use *must* + have + perfect infinitive.

Example:
· In October 1971, Joseph Hafele and Richard Keating conducted what must have been a **grueling** experiment.

grueling 厳しい

(C) To show impossibility, use *cannot / could not* + have + perfect infinitive.

Example:
· We could not have talked about gravitational potential energy and **elastic** potential energy at the same time.

elastic 弾性

There is ... construct:
· There is *virtually/almost* no doubt.
· There is a *high/strong probability/possibility*.
· There is a *slim/slight/remote* chance.

virtually 事実上

Text: Computers

Computers have forever definitely changed the way we live and work. They have transformed our ability to deal with information and data in particular. We are **swiftly** approaching the point where we may be able to analyze information **infinitely** quickly, store an unlimited quantity of data, and communicate data **instantly**. Computers have also enabled the common person to produce goods that formerly required the resources of major organizations. They are perfect for high-volume computing jobs including statistical and mathematical data computation and analysis, as well as scientific and engineering calculations.

swiftly 速く
infinitely 無限に
instantly すぐに

As technology advances, new computer models and types **emerge**. Hardware is at the heart of every computer. However, without the use of software, computers are merely boxes that are incapable of performing any computations or activities.

emerge 出現する
incapable ... or ⇒ 16.7

It is difficult to say which device was the first computer. Originally, the term "computer" referred to a person who used a mechanical calculating device to perform numerical calculations. The **abacus**, **slide rule**, and, <u>most likely</u>, the **astrolabe** and Antikythera mechanism (circa 150–100 BC) are examples of early mechanical computing machines. In 1801, Joseph Marie Jacquard improved the **textile loom** by using a set of **perforated** paper cards as a template to enable his loom to mechanically weave **intricate** designs. This can probably be considered as an early kind of programmability, however limited. Charles Babbage invented and built the first fully programmable mechanical computer, **dubbed** "The Analytical Engine," in 1837. Several technologies that <u>would</u> later be useful in the development of practical computers <u>had emerged</u> by the end of the nineteenth century, including the punched card, Boolean **algebra**, **vacuum tube** (**thermionic valve**), and **teleprinter**. In the early half of the twentieth century, more complex analog computers that used a direct mechanical or electrical representation of the problem as a basis for calculation could <u>meet</u> many scientific computing <u>demands</u>. Digital electronics (primarily invented by Claude Shannon in 1937) and more flexible programmability were significant advancements. The US Army's **Ballistics** Research Laboratory's ENIAC (1946) was the first <u>general-purpose electronic computer</u> to use **decimal arithmetic**. Throughout the 1950s, vacuum <u>tube-based computers</u> were used, but by the 1960s, <u>transistor</u>-based computers had largely replaced them because they were smaller, faster, cheaper, required less power, and were more reliable. Because of the Internet's widespread **proliferation** in the 1990s, nearly all modern electronic devices include some form of computer.

> Relevant vocabulary: applet, application, browser, **circuits**, CPU (central processing unit), data, database, desktop, email, graphics, keyboard, laptop, mainframe, monitor, mouse, notebook, operating system (OS), printer, program, RAM (random access memory), scanner, screen, search engine, server, software storage, spreadsheet, terminal, workstation

Side glossary:
most likely ⇒ G5.4
abacus そろばん
slide rule 計算尺
astrolabe 天体観測儀
textile loom 織機
perforated 穴のあいた
intricate 複雑な
dubbed ... ～と呼ばれる
would ⇒ G9.3
had emerged ⇒ G9.2.3
algebra 代数
vacuum tube 真空管（熱電弁）
thermionic valve 真空管
teleprinter 印刷電信機
meet ... demands ⇒ G16.5
ballistic 弾道
general-purpose electronic computer ⇒ G16.1
decimal arithmetic 十進法
tube-based computers ⇒ G16.1
transistor-based computers ⇒ G16.1
proliferation 増殖
circuit 回路

Exercises

Understanding the text
Q1 Write the definitions of the following parts of a computer:

CPU (Microprocessor), monitor, motherboard, power supply, primary storage (RAM)

Grammar point
Q2 Complete the following sentences by choosing a word according to the scale of probability:

Many scientists anticipate that fossil fuels will (a) _____ (100 %) **run out** by the middle of this century, whereas others say they will (b) _____ (75 %) run out sooner. Whatever the time scale, fossil fuels (c) _____ (100 %) will eventually come to an end, and we will need to investigate alternative energy sources. Alternative energy will be able to meet the world's demands in the short term (d) _____ (25 %), but our energy needs in the long term will (e) _____ (0 %) be met by fossil fuels.

run out 消費しつくされる

Writing
Q3 Write about three ongoing research efforts to improve computer speed and comment on the **likelihood** of the research succeeding using modal expressions of certainty and probability.

likelihood 可能性

第 25 章　確実性と可能性

研究について書く際，特に，確かな証拠がなくても強い可能性を述べなければならないことが

あります。一般的に可能性のあることがらについて述べることもあります。また，研究のある側面について確信がもてないのに，可能なはずだと思い込んでしまうこともあるかもしれません。100% 確実なことを述べることもあります。したがって，これらの異なる状況のどんな小さな違いをも表す助動詞を用いる必要があります。本章ではその表現について学びます。

25.1　確実性の度合い

確実性の度合いが 100% から 0% までの範囲であることを考慮することで，以下のカテゴリーに区別されます（数値は厳密ではありません）。

確実である（100%），可能性が高い（75%），可能性がある（50%），可能性が低い（10%），不可能である（0%）

表 25.1　各カテゴリーを表す表現

カテゴリー	例
確実である (100%)	・We are absolutely **certain/positive/sure** that torque and angular momentum are related. (トルクと角運動量の間に関係があることは間違いない。) ・The temperature is **unquestionably** higher than the Curie point. (温度は間違いなくキュリー点よりも高い。) ・The forces cannot cancel because one is **certainly/definitely** greater. (確かに／間違いなく力の片方が大きいので，力は相殺できない。)
可能性が高い (75%)	・The powder explosion will **most/very/quite** likely occur within the pipe. (パイプの内部で粉体爆発が生じる可能性は極めて高い。) ・This is the most **probable** distance between the single electron and the central proton. (この数値はおそらく 1 個の電子と中心部の陽子の距離を示しているだろう。) ・In a scientific experiment, increasing the concentration of the reactant **could** lead to a higher reaction rate. (科学実験では，反応物の濃度を上げると反応速度が上がる可能性がある。)
可能性がある (50%)	・The distance you perceive **may/might** not be the same as the true distance. (人間の知覚する距離は実際の距離とやや異なることがありうる。)
可能性が低い (10%)	・Momentum **should not** be confused with energy. (運動量とエネルギーを混同してはならない。) ・Slope measurement will **probably not** be precise. (傾斜の測定はおそらく正確ではないだろう。) ・It is **unlikely/improbable** that you will see upside-down racing. (逆さまのレースを見ることは考えにくい。)
不可能である (0%)	・A flick of the finger against the wall will **certainly/definitely** not cause damage. (指で壁をはじいても壁は決して傷つかない。) ・The intrinsic properties of the system **cannot possibly** change. (このシステムの固有の性質が変わることは考えられない。) ・In differentiating a vector product, be **sure/certain/positive** not to change the order of the two quantities. (ベクトル積を微分する際，決して 2 つの量の順序を変更してはならない。)

25.2　可能性に言及する条件法の文

接続詞 if は 2 通りの使い方があります。

〔1〕　**事象が可能性のある場合**

例：
・**If** an object **is** subjected to a force, then it **will** experience acceleration. (物質に力を加えると加速度が生じる。)

〔2〕　**事象がありえない場合**

例：
・**If** the lines **were** not parallel, an element arriving at their intersection **would** have two different velocities at the same time. (線が平行でない場合，それらの交点に到達する要素は，同時に 2 つの異

なる速度をもつ。)

25.3　可能性を表す表現

考察では，読者に議論をする機会を与えなければなりません。たとえ正論であったとしても，あまりにも確信に満ちた書き方は避けるべきで，確実性に幅をもたせた言い方をすべきです。仮定や確信の述べ方の例を以下に示します。

⑴　文頭または文末の表現：

We think/believe/guess/suppose/assume; In our opinion; We are certain/sure

⑵　関連するすべての事実を知っているわけではないとき，これまでに得た知見に基づき次のような表現ができる：

(A)　可能性に言及：may/ might/ could + have + 過去分詞

例：
・Some of you **may have** already **read** that chapter. (みなさんの中にはその章をすでにお読みになった方もおられるかもしれない。)

(B)　確信を述べる：must + have + 過去分詞

例：
・In October 1971, Joseph Hafele and Richard Keating carried out what **must have been** a grueling experiment. (1971 年の 10 月に，Joseph Hafele と Richard Keating は，困難と呼ぶほかない実験を行った。)

(C)　不可能：cannot / could not + have + 過去分詞

例：
・We could not have spoken of gravitational potential energy and elastic potential energy at the same time. (重力ポテンシャルエネルギーと弾性ポテンシャルエネルギーについて同時に述べることはできなかっただろう。)

> **There is ... 構文**
> ・There is *virtually/almost* no doubt. (事実上／ほとんど　疑いの余地はない。)
> ・There is a *high/strong probability/possibility*. (確率／可能性　は　高い／低い。)
> ・There is a *slim/slight/remote* chance. (可能性はわずかである。)

テキスト：コンピュータ

コンピュータは私たちの生活や仕事を根底から変えました。中でも，情報やデータの処理を飛躍的に拡大させました。無限の情報をたちどころに分析し，無限のデータを保存して，そのデータをすぐに伝達することが可能になるかもしれない時が急速に近づいています。コンピュータはまた，これまでは大手の企業や組織でしか生み出せないものを，誰もが作りだせるようにしました。統計や数学のデータ計算や分析，科学や工学の計算などといった大容量のコンピューティング・ジョブに，コンピュータは最適です。

技術の進歩に伴い，新しいコンピューター・モデルとタイプが登場しています。ハードウェアはすべてのコンピュータの心臓部ですが，ソフトウェアを使わないコンピュータは箱にすぎなくなり，計算やアクティビティを行うことはできません。

最初のコンピュータはどういうデバイスだったのか，よくわかっていません。コンピュータという語はもともと，機械的な計算装置を使って数値計算する人のことを指していました。そろばん，計算尺，それにおそらく天体観測儀と Antikythera 島の機械（紀元前 150-100）は初期の機械式計算装置の例です。1801 年に Joseph Marie Jacquard が織機を改良しました。穴の開いた紙のカードを型紙に用い，機械的に精巧なデザインを織機で織ることができるようにしたのです。その機能はきわめて限定されていましたが，これが最初のプログラム化された計算機と考えられます。1837 年に Charles Babbage がプログラム可能な機械式コンピュータ（"The Analytical Engine"）を考案しました。19 世紀の間に，パンチカードやブール代数，真空管，印刷電信機など，後のコンピュータの実用化に必要な技術がいくつか生み出されていました。20 世紀前半には，科学的な計算を求める声に，複雑度を増したアナログコンピュータが応えるようになりました。計算の基礎として，問題を直接，機械的あるいは電気的表現により処理したのです。1937 年に Claude Shannon らによって作られたデジタル回路と，一層

柔軟性を増したプログラミングの可能性は大きな前進でした。十進法を導入したアメリカ陸軍 Ballistics Research Laboratory の The ENIAC（1946）が，汎用性をもつ最初の電子コンピュータでした。1950 年代は真空管式コンピュータが使われていましたが，1960 年代にトランジスタ型に変わりました。小型であることや計算の速さ，値段の安さ，消費電力の少なさ，信頼度の高さが要因でした。1990 年代のインターネットの爆発的使用により，電子機器はほとんどすべてコンピュータを使用するようになりました。

コンピュータ用語：applet, application, browser, circuits, CPU（central processing unit）, data, database, desktop, email, graphics, keyboard, laptop, mainframe, monitor, mouse, notebook, operating system (OS), printer, program, RAM（random access memory）, scanner, screen, search engine, server, software storage, spreadsheet, terminal, workstation

Chapter 26

Coherence and cohesion

The consistency of an argument is established with the help of "coherence" and "cohesion." **Coherence** refers to the connection of ideas on a larger scale, whereas **cohesion** refers to the connection of sentences. A well-organized paper employs techniques to increase cohesion and coherence between and within paragraphs in order to guide the reader through the paper by connecting ideas, developing details, and strengthening the argument.

coherence [kəuˈhiərəns] 首尾一貫性 ⇒Ⓦ17
cohesion [kəuˈhiːʒn] 結束性 ⇒Ⓦ17
well-organized ⇒Ⓖ16.1

26.1 Difference between coherence and cohesion

The multiple ways in which pieces of a text are linked together (grammatical, **lexical**, **semantic**, **metrical**, **alliterative**) are referred to as cohesion. Cohesion varies from coherence **in that** a document may be cohesive but incoherent, i.e., make no sense. Coherence refers to the interconnection of concepts on a broader scale, whereas cohesion refers to the interconnection of sentences. Another way to think about it is that coherence refers to paragraphs, while cohesion refers to how phrases and sentences connect with each other.

lexical 語彙の
semantic 意味の
metrical 韻律の
alliterative 頭韻の
in that … 〜という点において ⇒Ⓖ7.1

26.2 Building coherence

Your writing must have coherence for the reader to understand it completely. The following are the steps needed to ensure that your text has coherence.

(1) Understanding and knowing what you want to say is the first step toward coherence.
(2) You must comprehend the overall structure of the form of writing, reading, or speaking in which you are involved, as well as the context.
(3) Between phrases and paragraphs, you must employ correct discourse markers/sentence connectors/links. You must also use complete sentences.

26.3 Cohesion through discourse markers

In a text, cohesion is achieved through lexis (words) or grammar. Using **synonyms** to refer to the same notion is one technique to provide cohesion. Discourse markers are **another** option. Discourse markers are used to divide ideas in writing and speech. Discourse markers also show how ideas are connected.
Some common discourse markers are:

synonym 同義語
another もう一つの

additionally, but, however, in addition, in conclusion, in summary, moreover, not only ... but also, on the contrary, on the other hand, such as, therefore, though, thus, to summarize, whereas

26.4 Function of discourse markers

The functions of discourse markers are quite varied, including contrast, **deduction**, example, addition and **summation**. Table 26.1 shows some of the functions and examples of discourse markers.

deduction 演繹 ⇒Ⓦ10
summation 要約

Table 26.1 A few common discourse markers

Discourse marker	Function	Example	
and	add information	These are licensed copies and may not be sold or transferred to **a third party**.	not ... or ⇒ 16.7 **a third party** 第三者
but	contrast	I tried, but I could not think of a single example.	
to summarize	summarize/conclude	To summarize, the **thermal efficiency** calculated here applies only to Carnot engines.	**thermal efficiency** 熱効果
therefore	reason/result/cause/effect	I think, therefore I go fast!	
for example	give an example	Most of us, for example, cannot directly lift an **automobile** but can do so with a **hydraulic jack**.	for example ⇒ 16.3 **automobile** 自動車 **hydraulic jack** 油圧ジャッキ

26.4.1 Difference between *but* and *however*

In contrast to *but*, which differentiates between ideas in the same sentence, *however* distinguishes between ideas in different sentences.

Examples:
· The block will no longer be **stationary** but will begin to move upward.
· The product of force and distance remains constant, resulting in the same amount of work being done. However, there is frequently a significant advantage to being able to **exert** greater force.

stationary 静止している

exert はたらかせる，行使する

26.4.2 Connecting ideas

There is a slight difference in how we use different discourse markers to connect ideas in our text.

In addition – Connect ideas in different sentences (followed by a **clause**)

clause 節

Example:
· The measured magnetic field may differ in magnitude and direction at any point on the Earth's surface. *In addition*, the field observed at any point on the surface of Earth changes **over time**.

over time 時間の経過とともに

And – Connect ideas in the same sentence (followed by a clause)

Example:
· **Magnetometers** measure angles and determine the magnetic field.

magnetometer 磁力計

As well as – Connect ideas in the same sentence (followed by a **noun group**)

noun group 名詞句

Example:
· Wavelengths, as well as angles, can be used to describe **phase** differences.

phase 位相

Text: Information technology (IT)

The use of technology to solve business or organizational **challenge**s on a large scale with the help of a computer system is known as information technology. In contrast to personal or recreational technology, Information Technology or IT is usually used in the context of commercial activities.

challenge 難題

Information systems collect, organize, store, analyze, **retrieve**, and display data in a variety of media (text, video, and speech). IT allows for the automated **manipulation** of

retrieve 取得する
manipulation 操作

Chapter 26

digital data as well as the **conversion** of digital data to and from analogue formats. **Integrated circuits** and digital communications have been the **driving forces** behind the creation of the essential hardware. Software has advanced at the same time, with easy-to-use solutions for creating, maintaining, manipulating, and **querying** files and records. Most of these software products **are intended for** both **experienced** computer users and enthusiastic **amateurs**.

The evolution of computer networks is another major element of IT. Techniques, physical connections, and computer programs used to connect two or more computers are all part of a network. Users on a network can share files, printers, and other resources, as well as send and receive e-mail and run programs on other computers. Each network follows a set of computer programs known as network **protocols** that allow computers to communicate with one another. Gateways have made it possible to connect computer networks more efficiently. The World Wide Web is the largest network. It is made up of many smaller interconnected networks known as internets. Tens, hundreds, or thousands of computers may be connected through these internets. They can share information, such as databases, with one another. People all around the world can connect with one another efficiently and inexpensively because of the internet.

> 💡 **Relevant IT vocabulary:** bandwidth, bps (bit per second), baud, **configure**, download, hack, hub, install, ISP (Internet service provider), LAN, **optical fiber**, packet, signal, switch, **transmission**, **transmit**, upload, WAN (wide area network), webpage, website, wireless

conversion 変換する
integrated circuit 集積回路
driving force 駆動力
easy–to–use ⇒ G16.1
query 問う，検索する
be intended for … 〜を対象としている
experienced 経験豊富な
amateurs [ˈæmətər] または [ˈæmətʃər] アマチュア，愛好家
another もう一つの
protocols プロトコル
bandwidth 帯域幅
bps ビット毎秒
baud [bɔːd] 通信回線が1秒当たりに送信可能な情報量の単位
configure（名詞）構成，設定
optical fiber 光ファイバ
transmission 送信
transmit 送信する
WAN 広域通信網
prefix 接頭辞
transaction 取引，情報のやりとり
transfer 転送する
compatible 互換性のある
configure（動詞）設定する

Table 26.2 Prefixes used in IT:

Prefix	Meaning	Example
Inter-	between	interactive, interconnect, internet, international
Intra-	within	intranet
Trans-	across	**transaction**, **transfer**, transmit
Co-/com-/con-	with	combine, **compatible**, **configure**, connect
Up-	To the internet	upload
Down-	From the internet	download, downtime

Exercises

Understanding the text

Q1 Complete the following sentences with words taken from the box in the text concerning "relevant IT vocabulary":

i. The speed of data transfer in a **communications system** is measured with _____.

ii. To send out a signal is to _____.

iii. _____ is a single file on the World Wide Web that contains text, graphics, etc. and is linked to other files on a website.

communications system
通信システム（通信の意味では communication は複数形。コミュニケーションの意味では単数形）

Grammar point

Q2 Prepare a table similar to Table 26.2 for the discourse markers:
on the other hand, however, on the contrary, and *thus*.

第 26 章　coference と cohesion

　議論の首尾一貫性は coherence と cohesion があれば保たれます。coherence は広くアイデア間の結びつきを指すことばであり，cohesion は文と文との間の結びつきを指すことばです。構成のしっかりした論文では，パラグラフ間やパラグラフ内の首尾一貫性や結束性を高めるテクニックを用い，アイデアを結びつけ，詳細を展開し，論旨を強化して，読者の理解を促進します。

26.1　coherence と cohesion の違い

　テキストを構成する個々の要素を互いに結びつける複数の方法（文法的，語彙的，意味的，比喩的，韻律的）を，cohesion と呼びます。cohesion は coherence とは異なります。文書は cohesion を保持していながら coherence を欠く，つまり意味が通じないことがあります。cohesion が文と文とのつながりを指すのに対し，coherence はもっと広い文脈での概念間のつながりを指します。別の考え方をすれば，coherence とはパラグラフとパラグラフの関係を指し，結束性は句と句，文と文とのつながりを指します。

26.2　coherent な文章の作成

　読者に正確に理解されるよう，文章には coherence をもたせなければなりません。以下，テキストにもたせる手順を示します。

(1)　自分の言いたいことかを理解して知ることが，coherence への第一歩である
(2)　書く・読む・話す際には，全体的な文章の構造や文脈を把握しなければならない
(3)　句と句の間，パラグラフとパラグラフの間に適切な discourse marker（談話マーカー）や文と文を繋ぐことばを用いなくてはいけない。また，文をきちんと書けなくてはいけない

26.3　discourse marker を用いた cohesion の保持

　cohesion を保証するのは語と文法です。例として同義語の使用があります。同じ意味のとき同義語を使うことで cohesion が保たれます。discourse marker の使用も有効的です。書くときも話すときも discourse marker は，一つのアイデアと別のアイデアを区別したり関連づけたりしてくれます。

　よく用いられる discourse marker を挙げます。

additionally, but, however, in addition, in conclusion, in summary, moreover, not only ... but also, on the contrary, on the other hand, such as, therefore, though, thus, to summarize, whereas

26.4　discourse marker の役割

　discourse marker の機能は多岐にわたります。対照，演繹，例示，付加，要約などです。以下の表 26.1 は discouse marker の機能と例をいくつか挙げています。

表 26.1　広く用いられる discourse marker

discourse marker	機能	例
and	情報の付加	These are licensed copies **and** may not be sold or transferred to a third party.（これらの文書にはライセンス権が与えられており，第三者に売ったり提供したりすることはできない。）
but	対照	I tried, **but** I could not think of a single example.（例を考えに考えたが，一つも思い浮かばなかった。）
to summarize	要約 / 結論	**To summarize**, the thermal efficiency calculated here applies only to Carnot engines.（まとめると，ここで計算した熱効果は Carnot エンジンにしか当てはまらない。）
therefore	理由 / 結果 / 原因 / 結果	I think, **therefore** I go fast!（我思う，ゆえに我進歩す。）
for example	例示	Most of us, **for example**, cannot lift an automobile directly but can with a hydraulic jack.（たとえば，自動車を人力で持ち上

| | | げることはほぼ不可能だが，油圧ジャッキを使えば可能だ。） |

26.4.1　接続詞 but と however の違い

but は同じ文の中のアイデアを区別するの対し，however は異なる文の中のアイデアを区別します。

例：
・The block will no longer be stationary **but** will begin to move upward. （この塊は静止することはなく，上方に動き出す。）
・The product of force and distance remains constant, resulting in the same amount of work being done. **However**, there is frequently a significant advantage to being able to exert greater force. （力と距離の積は一定のままであるため，同じ量の仕事が実行される。ただし，より大きな力を発揮できることには，多くの場合，大きな利点がある。）

26.4.2　アイデアとアイデアを結びつける

テキスト内のアイデアを結び付けるさまざまな discouse marker の用い方には，わずかな違いがあります。

in addition – 異なる文を結びつける（節が後続する）

例：
・The measured magnetic field may differ in magnitude and direction at any point on the Earth's surface. **In addition**, the field observed at any point on the surface of Earth changes over time. （測定された磁場は，地球の表面のどの点でも大きさと方向が異なる場合がある。さらに，地球の表面の任意の点で観測された磁場は，時間の経過とともに変化する。）

and – 同じ文の中にあるアイデアを結びつける（節が後続する）

例：
・Magnetometers measure angles **and** determine the magnetic field. （磁力計は角度を測り磁場を決定する。）

as well as – 同じ文の中にあるアイデアを結びつける（名詞句が後続する）

例：
・Wavelengths, **as well as** angles, can be used to describe phase differences. （波長と角度の両方を使用して位相差を表すことができる。）

テキスト：情報技術（IT）

　私たちはコンピュータシステムを用いて，ビジネスや組織の抱える大きな課題の解決を図ります。このコンピュータシステムのはたらきを情報技術と呼びます。個人的な，または娯楽的な技術とは対照的に，情報技術もしくは IT（Information Technology）ということばは普通，ビジネスの文脈で用いられます。

　情報システムは，さまざまなメディア（テキスト，ビデオ，および音声）のデータを収集，整理，保存，分析，取得，表示します。IT はデジタルデータを自動的に操作し，デジタルからアナログへ，またアナログからデジタルへ変換します。集積回路とデジタル通信が，コンピュータの重要なハードウェアを生みだす原動力になっています。同時にソフトウェアも開発が進み，ファイルやレコードを簡単に作成，保存，操作，検索できるようになっています。このようなすぐれたソフトウェア製品のほとんどは，コンピュータを専門に扱う人やその熱心な愛好家に向けに作られています。

　また IT の大きな特色として，コンピュータネットワークの進化が挙げられます。2台以上のコンピュータを接続するのに使う技術や物理的接続，コンピュータ・プログラムはすべてネットワークの一部です。ネットワークの利用者は，E メールを送受信し，他のコンピュータ上でプログラムを動かせるのに加え，ファイルやプリンターなどを共有できます。それぞれのコンピュータは，ネットワークプロトコルというコンピュータプログラムに従い，コンピュータ間で通信を行います。ゲートウェイはコンピュータネットワーク間の接続を容易にしています。最大のネットワークが World Wide Web です。Wordl Wide Web はインターネットとして知られる多くの小さな相互接続されたネットワークからすべて構成されます。これら

のネットワークにより何十，何百，何千のコンピュータを接続でき，コンピュータ間でデータベースのような情報を共有できるのです。インターネットのおかげで，世界中の人々が，簡単に，そして費用をかけることなく結ばれています。

IT 関連の語：bandwidth, bps, baud, configure, download, hack, hub, install, ISP, LAN, optical fiber, packet, signal, switch, transmission, transmit, upload, WAN, webpage, website, wireless

---

表 26.2　IT で使われる接頭辞

接頭辞	意味	例
inter-	between	interactive, interconnect, internet, international
intra-	within	intranet
trans-	across	transaction, transfer, transmit
co-/com-/con-	with	combine, compatible, configure, connect
up-	to the internet	upload
down-	from the internet	download, downtime（休止時間）

Chapter 27

Exemplifying

exemplifying 例示する

The best way to clarify a point after giving a definition or making a general statement is to give an example of it. Transforming a **large** or abstract idea to a specific or concrete image is the function of an example. The impact of the example is that it makes the **assertion** more memorable, **intriguing**, and **persuasive** while also giving evidence for it. Scientists use examples to explain and clarify concepts, as well as to provide evidence for **corroboration**. Debating the **validity** of a concept with examples may be quite useful. If you can find an example to **demonstrate** a point, the meaning of your point will certainly become clearer.

large 重要な
assertion 主張
intriguing 興味深い
persuasive 説得力のある
corroboration 確証
validity 妥当性
demonstrate 具体的に示す

27.1 Language of exemplification

A good example of how to make an impression with a few words, a sentence, or a paragraph, needs to be studied carefully. Examples are sometimes not indicated with **clue** words, such as *for example, for instance*, and *to illustrate*; instead, the context needs to be understood to identify the example being provided. To **clarify** the many aspects of a concept, more than one example is frequently used.

clue 手がかり

clarify 明らかにする

The topic statement of a paragraph can be made clear and memorable by using examples. Indeed, examples enable readers to comprehend the topic sentence even when they <u>cannot</u> understand the meaning of <u>some</u> words.

not ... some ⇒ⓖ3.5

If you are giving multiple examples to support a single point, arrange them **in ascending order** of importance. This will strengthen your writing because the strongest example will be in the section with the greatest impact.

in ascending order 昇順で

27.1.1 Sentence patterns using exemplifying words and phrases

Some common exemplifying words and phrases are:

*for example, for instance, **to be specific**, to illustrate*

The position of these phrases can **vary** in a sentence.

to be specific 具体的に言うと
vary 変わる

(1) At the beginning of a sentence

Examples:
- For example, if the charge is <u>**uniformly**</u> distributed over a **sphere**, we can **enclose** the sphere in a **spherical** Gaussian surface.
- For instance, the figure **depicts** a particle with a **position vector**.
- To be specific/Specifically, the <u>standard for</u> the meter was redefined to be 1 650 763.73 wavelengths of a particular <u>orange-red light</u> emitted by krypton-86 atoms.

uniformly 均一に ⇒ⓖ5.4
sphere 球
enclose 囲む
spherical [ˈsfɪrɪkl] あるいは [ˈsferɪkl] 球状の
depict 記述する
position vector 位置ベクトル
standard for ⇒ⓖ7.5
orange-red light ⇒ⓖ16.1

(2) In the middle of a sentence

Example:
- <u>**Assume**</u> you start with an uncharged **capacitor**, <u>for example</u>, and remove electrons from one plate.

assume ... ～と仮定しよう ⇒ⓦ16
capacitor コンデンサ
for example ⇒ⓖ16.3
exemplification 例示

27.1.2 Change of word forms to show exemplification

The same word can be used as a noun or a verb in order to provide an example. When writing, you should keep in mind that examples are frequently written about using the present simple tense.

(1) Using nouns:

Examples:
- A magnetic **dipole** is something that has two **poles**, an example of which is a magnet.
- In such a case, shock waves result.
- As an instance of <u>his method, Bacon</u> describes <u>an experiment into</u> the **nature** and cause of the rainbow.
- It is quite clearly described with a diagram, as an illustration of the phenomena of vision.

dipole 多極子
pole 極
<u>his method, Bacon</u> ⇒ⓖ3.9
<u>an experiment into</u> ⇒ⓖ7.5
nature 性質

(2) Using verbs:

Examples:
- The trajectory of a projectile launched into the air exemplifies the principles of Newton's laws of motion.
- The figure illustrates this fact.
- This may be exemplified in the case of **alkalimetry**.
- This relation is illustrated by the phasor diagram.

alkalimetry アルカリ滴定

(3) *Such as* and *like*:

Examples:
- Instruments that measure **alternating current**, such as ammeters and voltmeters, are usually calibrated.
- We are talking about the fundamental properties of an electron, like its mass and electric charge.

alternating current 交流電流

Text: Interactive multimedia

Text and visuals are combined with digitized sound and music in **interactive** multimedia. Computers can **handle** music files and full-motion video images as **readily** as they can manage text files. The computers can thus display movie-quality images while playing music in high-fidelity digital audio stereo. **In a nutshell**, multimedia is an ideal balance of print, audio and video news. It has become **conceivable** because of quickly changing digital technology and the efficiency of computers in **manipulating**, **storing**, and **retrieving** data.

Multimedia can refer to a variety of things such as a multimedia electronic mail transmitted over the internet or a CD-ROM* encyclopedia. **Indisputably**, text has become a more powerful force as a result of multimedia software. There are many advantages for interactive multimedia. For instance, despite the fact that no **comprehensive** study on the impact of interactive multimedia learning exists, some say that interactive technology helps students learn faster. This is because <u>self-paced</u> individualized teaching, as well as immediate interaction and feedback, is available with multimedia.

In summary, the flexibility to explore information in whatever way is most meaningful for individual users is one of the major characteristics of interactive multimedia.

___
* **CD-ROM** Compact Disc. read only memory

interactive 双方向の
handle 処理する
readily 迅速に
in a nutshell 手短にいうと
conceivable 考えられる
manipulate 操作する
store 保存する
retrieve 取得する
<u>**indisputably**</u> 間違いなく ⇒ⓖ5.3
comprehensive 包括的な

<u>self-paced</u> ⇒ⓖ16.1

Exercises

Understanding the text
Q1 Why are interactive multimedia applications important?

> **Grammar point**
> **Q2** Write a sentence giving an example of <u>an energy-saving innovation</u>.
>
> **Writing**
> **Q3** Write a paragraph discussing how interactive learning may or may not influence students' learning. Give one or two examples to clarify your meaning.

an energy-saving innovation
⇒ G 16.1

第 27 章　例示する

　定義を与え，あるいは一般的なことを述べた後にそれらを明確にするのに最善の方法は，例を挙げることです。一般的あるいは抽象的なアイデアを，特定の具体的なイメージに転換させるのが，例の役割です。例の大切さは，主張がより記憶に残り，興味深く，説得力のあるものになると同時に，主張の証拠を提供するところにあります。科学者は例を用いて概念を明確に説明し，確固たる証拠を提示します。実例を用いて概念の妥当性を論じることは，とても役に立つかもしれません。主張したいことを具体的に示す例を見つけることができれば，その主張は確実に明確になります。

27.1　例示表現

　数単語や一文，または 1 パラグラフで読者に印象を与えるよい例示については，注意深く学ぶ必要があります．such as, for example, for instance, to illustrate といった手がかりの言葉で例が示されない場合があります。そのときには，文脈を手がかりに例が示されていることを判断しなければなりません。概念のもつ多くの側面を明確にするために，例を複数示すことがよくあります。

　例を用いることでパラグラフのトピック・センテンスを明確で印象的なものにすることができます。実際，読者は単語の意味が理解できなくても，例を通してトピック・センテンスを理解することができます。

　一つの論点を支持するのに複数の例を挙げる際には，重要度の高い順に並べてください。そうすれば，一番強い例が一番インパクトのあるセクションに置かれ，文章が強化されます。

27.1.1　例示語句を使った文のパターン

　例示を示す単語やフレーズ（句）に以下があります。

for example, for instance, to be specific, to illustrate

その位置は一様ではありません。

(1) 文頭：

例：

・ **For example**, if the charge is uniformly distributed over a sphere, we can enclose the sphere in a spherical Gaussian surface. (たとえば，充電が球体全体に一様に広がれば，球を球面ガウス面で囲むことができる。)
・ **For instance**, the figure shows a particle with position vector. (たとえば，この図は位置ベクトルをもつ粒子を示している。)
・ **To be specific/Specifically**, the standard for the meter was redefined to be 1 650 763.73 wavelengths of a particular orange-red light emitted by atoms of krypton-86. (具体的にいえば，リプトン 86 原子から発せられる燈赤色光の波長 1 650 763.73 を，メートルの基準と再定義した。)

(2) 文中：

例：

・ Assume you start with an uncharged capacitor, **for example**, and remove electrons from one plate. (たとえば，充電されていないコンデンサから始めて，一つのプレートから電子を除去するとしよう。)

27.1.2　異なる品詞を用いた例示表現

　1 つの例示語を名詞と動詞で用いることができます。論文やレポートを書く際に注意しなければならないのは，例示は現在形を用いて記すことが多いということです。

(1) 名詞を使う

例：
- A magnetic dipole is something that has two poles, **an example** of which is a magnet. (磁気双極子とは2つの極をもつもので，その例として磁石がある。)
- In such **a case**, shock waves result. (その場合，衝撃波が生じる。)
- As **an instance** of his method, Bacon describes an experiment into the nature and cause of the rainbow. (その手法の例として，Bacon は虹の性質と虹が生じる原因に関する実験について説明している。)
- It is quite clearly described with a diagram, as **an illustration** of the phenomena of vision. (視覚現象の説明として，それはダイアグラムを用いて明確に記述されている。)

(2) 動詞を使う

例：
- The trajectory of a projectile launched into the air **exemplifies** the principles of Newton's laws of motion. (空中に発射された発射体の描く軌跡は，ニュートンの運動法則の原理を例証している。)
- The figure **illustrates** this fact. (この図はこの事実を示している。)
- This may **be exemplified** in the case of alkalimetry. (これの例としてはアルカリ滴定がある。)
- This relation **is illustrated** by the phasor diagram. (この関係はフェーザ図で示される。)

(3) such や like

例：
- Instruments that measure alternating current, **such as** ammeters and voltmeters, are usually calibrated. (電流計や電圧計などの交流電流を測定する機器は，通常，較正されている。)
- We are talking about the fundamental properties of an electron, **like** its mass and electric charge. (質量や電荷など，電子の基本的な性質について話している。)

テキスト：双方向性マルチメディア

　双方向性マルチメディアでは，テキストと視覚情報がデジタルサウンドや音楽と協働します。コンピュータはテキストファイルを処理するのと同じように手軽に，音楽ファイルやフルモーションビデオの画像を処理できます。こうしてコンピュータは，音楽を高音質のデジタル・オーディオ・ステレオで再生しながら，映画と同じような鮮明な画像を写しだすことができます。要するにマルチメディアは，紙媒体の音声ニュース，ビデオニュースの理想的なバランスです。それは，デジタル技術の急速な変化と，コンピュータによる効率的なデータの操作と保存，取得のおかげで考えられるようになりました。

　マルチメディアには，インターネット上で送信するマルチメディア電子メールや CD-ROM 版百科事典などさまざまなものがあります。マルチメディアソフトウェアの出現でテキストの威力は間違いなく大いに増しました。双方向性マルチメディアには多くの利点があります。たとえば，双方向性マルチメディアを活用した学習の効果についての包括的な研究はなされてはいませんが，双方向性技術が学習の効率を高めているとの指摘をする人もいます。マルチメディアにより自分のペースで個人指導ができ，反応やフィードバックがすぐに可能になるからです。

　つまり，個々の使用者にとって最も意味のあるやり方で情報を得ることができるという柔軟性が，双方向性マルチメディアの大きな魅力の一つです。

Chapter 28

Explaining future plans

Future plans include both short-term and long-term objectives. A good way to prepare for explaining these plans is to think about what you want to do in the next five years. You can even go a step further and plan for the next **decade**.

decade 10 年

28.1　Language expressing future plans

The conclusion of an article is the place at which a suggestion is made for the future. This suggestion may **imply** either certainty, probability or possibility.

imply 暗示する

> 💡 **Difference between *probable* and *possible*:** The term *probable* implies that a given **incident** has a very high chance of occurring. "Possible", on the other hand, **denote**s that an incident may or may not occur, and there is no surety of the outcome.
>
> Examples:
> · It is probable that the increase in greenhouse gas emissions is contributing to global warming.
> · It is possible that a new drug could cure COVID-19.

incident 出来事
denote 意味する

(1) The future simple tense is formed with the help of an **auxiliary verb** *will* and the **infinitive** form of the main verb. *Will* is one of the most established ways to discuss the future.

auxiliary verb 助動詞
infinitive 不定詞の

> Example:
> · In physics problems, the student will learn how to recognize common patterns.

(2) To announce your plans, the present perfect tense can also be used.

> Example:
> · We have decided to do some further experiments.

(3) We can **utilize** the present continuous tense to discuss future arrangements particularly when we have agreed on a time and place with others. A frequently used expression is "be going to."

utilize 利用する

> Examples:
> · The students are meeting their professor tomorrow.
> · The students are going to meet their professor tomorrow.

The verbs *plan* and *think* can be used in the present continuous to communicate about tasks that are not completely certain. *To* is used after words such as *plan, intend, want, hope.*

> Examples:
> · We are planning to divide the disk into **concentric** flat rings.
> · The quiz questions are intended to be **straightforward**.
> · We want to explore the mysteries of the universe through scientific inquiry.
> · We hope to produce a narrower beam by using narrower slits.

concentric 同心の
straightforward 率直な,
簡単な

(4) Present simple tense is also used to indicate the future state and specifies actions that will occur according to schedule, especially with the verbs: arrive, begin, come, depart, end, go, leave, start, end.

> Example:

· The light waves arrive at a common point in all situations.

Instead of future tense forms, the verb forms of the present simple tense are used in **subordinate clauses** of time and conditions.

subordinate clause 従属節

Example:
· I will start a new project after I end this one.

(5) The future perfect tense is used to forecast that an action or event will be completed before (or by) a certain date in the future.

Example:
· When the light waves leave the two media, they will have converged on the same wavelength.

converge 収束する

(6) *Can, must, have to*, and *need to* are common words that apply to both the present and the future. You can use *will be able to, will have to*, and *will need to* to emphasize that they refer especially to the future.

to both A and B ⇒🄖7.4

especially ⇒🄖5.5

Examples:
· The **capacitor** will be able to **withstand** a large potential difference.
· You will need to convert radians to degrees in the calculations.

capacitor コンデンサ
withstand 抵抗する

28.2 Predictions, assumptions, guesses about the future

Future predictions, assumptions, and guesses are often accompanied by introductory combinations *I think, I suppose, I guess, I'm sure*, *probably,* etc.

prediction 予測
assumption 仮定
guess 推測

Examples:
· You are probably good at driving a car on a highway.
· We suppose that every measurable **property** will <u>assume</u> a stable, unchanging value.
· We can guess what the answers must be.
· We commonly think of these particles as the **building blocks** of **matter**.
· We can be certain that the **net** force acting on the particle is constant in magnitude if it moves in a circle at constant speed.

property 性質
assume 取る⇒🆆16
building block 構成要素
matter 物質
net 正味の

28.3 Conditionals

conditional 条件文

Conditionals are used when we are talking about what will happen in the future under certain conditions. The first conditional is used to discuss events that are expected to occur in the future under specific real or hypothetical circumstances. For the first conditional, the main clause uses the future tense, but the *if* clause uses the present tense.

Example:
· If the two **currents** are antiparallel, the two wires will <u>repel</u> each other.

current 電流 ⇒🆆4
repel 反発する ⇒🆆7
subordinate clause 従属節

This rule also applies to conditional sentences with the conjunction and **subordinate clauses**:

unless, when, as soon as, before, after, until.

Examples:
· Unless the loop is **superconducting**, energy will always be transferred to thermal energy during the process.
· It will accelerate until it reaches a certain terminal speed.
· The atoms will **collide** as soon as a magnetic dipole moment **aligns** with the external field.
· When a sample's temperature exceeds its Curie temperature, **exchange coupling** will disappear.
· After being reflected by a mirror, only **rays** that are relatively close together will enter

superconducting 超電導な

collide 衝突する
align [əˈlaɪn] 一直線に並ぶ
exchange coupling 交換相互作用
ray 光線

the eye.

· It will only last for a short time before **decaying**.

decay 腐敗する

The second conditional is employed to communicate about situations that are unlikely or impossible to happen in the future. The past simple in the condition refers to the present moment.

Example:

· My car would not emit any carbon into the atmosphere if it had a **hydrogen fuel cell**.

hydrogen fuel cell 水素燃料
電池
be bound to 必ず

Use *when* when you are sure about what you are talking about, i.e., it **is bound to** happen; when talking about things that might happen use *if*.

Example:

· When water reaches its boiling point at 100 degrees Celsius, it undergoes a phase change from liquid to vapor.

· If the temperature increases, the reaction rate of this chemical process may also increase.

Text: Virtual reality (VR)

A computer-generated simulation in which a human can interact within an artificial three-dimensional environment is referred to as virtual reality (VR). The person enters this virtual world or **is immersed** in this environment and can **manipulate** things or perform **a sequence of** actions while there. Virtual reality is a cybernetic experience that can be both **comparable** and dissimilar **to** the actual world. A three-dimensional (3D) movie is probably the most basic example of virtual reality. The head-mounted display (HMD) is the most immediately recognizable component of VR. Traditional virtual reality technologies use visual and audio stimuli to generate human experiences. **Augmented reality** and **mixed reality** are two other sorts of modern VR-style technology. Virtual reality has a variety of uses, including entertainment (such as video games), education (such as medical or military training), and business (e.g., virtual meetings). It is a wonderful method that relies only on technological advancements. These developments have certainly provided useful tools for medical and behavioral studies.

Virtual reality is being thought of as providing a lot of opportunities for the future as a consequence of the COVID-19 pandemic. With new virtual meeting room services, many companies are hoping to realize great profits. Virtual reality in education is on the horizon, and it will **undoubtedly** alter the way we live. Classrooms in the twenty-first century will be technologically upgraded, with virtual reality technology **considerably enhancing** student **engagement** and learning.

A computer-generated simulation ⇒ⓖ16.1
is immersed 浸っている
manipulate 操作する
a sequence of 一連の
comparable to 匹敵する
a three-dimensional 3D movie ⇒ⓖ16.1
the head-mounted display (HMD) ⇒ⓖ16.1
augmented reality 拡張現実
mixed reality 複合現実
VR-style technology ⇒ⓖ16.1
e.g. ⇒ⓖ16.9.6

be on the horizon 始まったばかり
undoubtedly 間違いなく ⇒ⓖ5.4
the way we live 人間の生き方
with ⇒ⓖ7.2
considerably かなり
enhance 高める
engagement 関与, 参加

Exercises

Understanding the text

Q1 Rewrite the last paragraph of the article on *Virtual reality* in this chapter using words and phrases denoting the future found in Section 28.1.

Grammar point

Q2 Rewrite the following sentences using the words in brackets with the future perfect tense.

 i. They will almost certainly increase funding for robotic exploration of the solar system. (likely)

 ii. It is likely that low-Earth **orbit** space travel will become more inexpensive. (possibly)

 iii. Scientists are unlikely to construct new <u>heavy-lift rockets</u> in the near future. (probably)

orbit 軌道

<u>heavy-lift rockets</u> ⇒ 16.1

Q3 Write sentences using the future perfect of the verb in brackets and *by, to* or *at*.

 i. CO_2 emissions are expected to increase by at least 21% by 2040, according to projections. (rise)

 ii. CO_2 concentrations are projected to reach 450 ppm by the year 2040. (increase)

Writing

Q4 Select a future possible development in your chosen subject of study. Write about how it may impact future research.

第 28 章　将来の計画の説明

　将来の計画には短期と長期の目的があります。これらの計画を説明しようとするとき，これから 5 年間にしたいことを思い浮かべてはどうでしょうか。さらに一歩進んで，その次の 10 年のことも考えてみてもよいと思います。

28.1　将来の計画を表現することば

　論文の結論では将来への示唆がなされます。この示唆には，実現可能性は，ほぼ実現しそうな場合（certainty）と可能性が高い場合（probability），可能性がある場合（possiblity）のいずれかが含まれます。

---

probable と possible の違い：

probable は出来事が起こる確率がとても高い場合に用いられるのに対し，possible は，出来事が実際に起きるかもしれないし，起きないかもしれない，また，結果がどうなるかはわからない，というときに用いられます。

例：

・It is **probable** that the increase in greenhouse gas emissions is contributing to global warming. (温室効果ガスの増加が地球の温暖化の大きな要因となっている可能性が高い。)

・It is **possible** that a new drug could cure COVID-19. (新薬が開発されれば新型コロナウイルス感染症の治療薬となる可能性はある。)

---

(1)　未来形は「助動詞 will ＋ 本動詞の原形」で表されます。will は未来を表す代表的な表現手段の一つです。

 例：

 ・In physics problems, the student **will** learn how to recognize common patterns. (物理の問題を解くことで，学生は一般的なパターンを認識する方法を学ぶ。)

(2)　計画を伝えるのに現在完了形も使います。

 例：

 ・We **have decided** to do some further experiments. (実験を重ねることにした。)

(3)　現在進行形を用いて先のことを表すことができます。特に，実施する時と場所が関係者間で決まっているような場合です。よく用いられる表現は be going to です。

 例：

 ・The students **are meeting** their professor tomorrow. (学生らは明日教授に会う。)

 ・The students **are going to** meet their professor tomorrow. (学生らは明日教授に会う。)

　plan と think を現在進行形で用い，完全には確定していないタスクについて伝えることがで

きます。plan や intend, want, hope などの動詞の後には to を補います。

例：
- We **plan to** divide the disk into concentric flat rings.（ディスクを分割して同心円状のフラットリングに分割する予定である。）
- The quiz questions **are intended to** be relatively straightforward.（クイズの問いは簡単なものにする予定だ。）
- We **want to** explore the mysteries of the universe through scientific inquiry.（科学的探究を通して私たちは宇宙の謎に迫りたいと思う。）
- We **hope to** produce a narrower beam by using narrower slits.（より狭いスリットを用いてより狭いビームを生成したい。）

(4)　現在形は未来の状態を示すためにも使われ，予定通りに起こる行動を述べます。特に arrive, begin, come, depart, end, go, leave, start といった動詞を用います。

例：
- The light waves **arrive at** a common point in all situations.（いかなる条件下でも光の波は同じ地点に達する。）

時間や条件を表す従属節では未来の出来事は未来形ではなく現在形で表されます。

例：
- I will start a new project after I **end** this one.（このプロジェクトが終わり次第，次のプロジェクトに取りかかる。）

(5)　未来完了形を用いて未来のある地点には完了しているであろうことがらを表現します。

例：
- When the light waves leave the two media, they will **have converged** on the same wavelength.（光の波が2つの媒体を離れるときには，同じ波長に収束するだろう。）

(6)　can, must, have to, need to は現在と未来のいずれにも言及します。未来であることをはっきりさせる場合には will be able to, will have to, will need to を用いることもできます。

例：
- The capacitor **will be able to** withstand a large potential difference.（コンデンサは大きな電位差に耐えることができる。）
- You **will need to** convert radians to degrees in the calculations.（計算する際，ラジアンを度に変換する必要がある。）

28.2　未来の予測，仮定，推測

未来の予測や仮定，推測は I think, I suppose, I guess, I'm sure, probably などで表現されることがよくあります。

例：
- You are **probably** good at driving a car on a highway.（君はおそらく，高速道路での運転は得意なのだと思う。）
- We **suppose** that every measurable property will assume a stable, unchanging value.（すべての計測可能な性質は安定していて，一定の値をもつものと仮定する。）
- We can **guess** what the answers must be.（解答がどのようなものか見当がつく。）
- We commonly **think** of these particles as the building blocks of matter.（これらの粒子はふつう物質の構成要素と考えられている。）
- We can be **certain** that the net force acting on the particle is constant in magnitude if it moves in a circle at constant speed.（粒子が円の中を一定の速度で運動していれば，その粒子にはたらく正味の力も一定であると考えてよい。）

28.3　条件文

一定の条件下で将来起こりうることを表現するのに条件文を用います。1つ目の条件文は，特定の現実または仮想の状況下で，将来発生すると予想される事象を表します。1つ目の条件文では，主節では未来形が，従属節では現在形が用いられます。

例：
- If the two currents **are antiparallel**, the two wires **will repel** each other.（2つの電流が逆平行なら，その電流が流れている電線は反発しあう。）

この規則は接続詞と従属節を伴う条件文にも適用されます。

unless, when, as soon as, before, after, unitil

例：

・**Unless** the loop is superconducting, energy will always be transferred to thermal energy during the process. （ループが超伝導でない限り，その過程でエネルギーは常に熱エネルギーに変換される。）

・It will accelerate **until** it reaches a certain terminal speed. （一定の終速度に達するまで加速は続く。）

・The atoms will collide **as soon as** a magnetic dipole moment aligns with the external field. （磁気双極子モーメントが外部磁場と整列すると同時に，原子は衝突する。）

・**When** a sample's temperature exceeds its Curie temperature, exchange coupling will disappear. （サンプルの温度がキュリー温度を超えると，交換相互作用はなくなる。）

・**After** being reflected by a mirror, only rays that are relatively close together will enter the eye. （鏡に反射した後，比較的近くに位置する光線のみが視覚にとらえられる。）

・It will only last for a short time **before** decaying. （しばらくすると腐敗する。）

2つ目のタイプの条件文は，めったに起こらない，あるいは起こりえない事態を表すのに用いられます。条件の表現に過去形が使われていれば，それは現在の事態に言及しています。

例：

・My car **would not** emit any carbon into the atmosphere **if** it **had** a hydrogen fuel cell. （水素燃料電池の車は炭素を排出しない。）

話している内容を確信しているとき，すなわち，確実に起きると思っているときには when を使い，起こるかもしれない事態を表す時には if を用います。

例：

・**When** water reaches its boiling point at 100 degrees Celsius, it undergoes a phase change from liquid to vapor. （摂氏 100 度の沸点に達すると，水は液体から蒸気に変化する。）

・**If** the temperature increases, the reaction rate of this chemical process may also increase. （温度が上昇すると，この化学プロセスの反応速度も増加する可能性がある。）

テキスト：仮想現実（VR）

　人工の 3 次元の環境の中で人間が交流するコンピュータが作り出すシミュレーションを仮想現実（VR）と呼びます。この仮想の世界に入り込み環境に浸ることで，人間はものを操作したり，一連の動作を行ったりすることができます。VR とは，現実世界と匹敵することも異なることもできるサイバネティックスな体験です。3 次元（3D）映画は最も基礎的な VR の例でしょう。頭部装着ディスプレイ（HMD）が最もわかりやすい VR の例です。これまでの VR 技術は，視覚と聴覚を刺激し，実際に経験したかのような間隔を人間にもたらすのに対し，最新の VR 技術は，拡張現実や「複合現実」を生み出しています。

　VR の利用は娯楽（たとえばビデオゲーム）や教育（医療，軍事訓練など），ビジネス（バーチャル会議など）と多岐にわたります。VR は科学技術の進歩に依存するすばらしい方法です。科学技術の進歩により，その進歩が医学の研究や人間の行動の研究を確かに前進させたのです。

　新型コロナウイルス感染症（COVID-19）が世界で猛威を振るう中，VR が将来にさまざまな機会を与えるものとして見なされつつあります。新しいバーチャル会議サービスを使い，多くの企業は大きな利益を得られるだろうと期待しています。教育での VR の活用は始まったばかりであり，きっと人間の生き方を変えるでしょう。21 世紀の教室では IT 技術が進歩し，VR 技術が生徒・学生の授業への積極的な関わりや学習を飛躍的に前進させるでしょう。

Making suggestions

When you make a suggestion, you are **putting forward** a plan or an idea for consideration by another person. Suggestions for future research have to be included in your research paper or presentation.

put forward 提示する

29.1 Language expressing suggestions

Making decisions **necessitates** a **thorough** examination of **alternatives**. A recommendation report considers a number of possible solutions to a problem before suggesting the best one. **Recommendation reports** are more **overtly** <u>compelling</u> because they provide a final recommendation based on study and data.

When the word *must* is used, it usually means that the speaker has decided that something has to be done. When the words *have to* is used, it usually means that someone else has decided that the activity is required.

necessitate 必要とする
thorough [ˈθɜːrəʊ] 徹底的な
alternative 他の選択肢
recommendation report 提案書
overtly [əʊˈvɜːrtli] あるいは [ˈəʊvɜːrtli] 明確に
compelling 説得力がある ⇒Ｗ7

Examples:
· The electric field must be defined for all points in the region.
· The particle itself does not have to rotate around the center.

Should is used to show that you are suggesting or **advocating** something. In the recommendations section of a paper or report, *should* is used with the passive tense.

advocate 提唱する（ad- = toward voc- = voice ～に向かい声を上げる）

Examples:
· During the parachute landing, the legs should be together.
· This is the maximum amount of stress **to** which the steel wires should be **subjected**.

subject A to B [səbˈdʒekt] A を B の下に置く ⇒Ｗ13

Combining a suggestive verb or an important adjective with the action you want the other person to take strengthens the suggestion. Rather than giving the person a command, the importance of the advice or recommendation is being emphasized.

Examples:
· I suggest that you try improving the engine efficiency.
· I recommend that you **extract** the energy as heat from the water.

extract とりだす ⇒Ｗ12

29.2 Ways of making a suggestion

Suggestions can be made in a variety of ways. The folllowing phrases may be useful if your recommendation is for yourself and others in a group:

· Why don't we do ... ?
· Let's try doing ...
· I would suggest that we do ...
· Let's do .. .
· What/How about doing ... ?
· We could do ...

Questions requesting suggestions include:

· What should I (or should we) do?
· What do you think I should do?
· What do you suggest we do?

29.2.1 Suggesting alternatives

When suggesting other options, the following phrases may be useful as it is essential to be extremely polite and not too **assertive**.

assertive 主張する

- How about ... ?
- (I think/I believe) it might help to ...
- It might be useful to ...
- I would suggest ...
- It'd be better to ...
- One thing you could do is work on ... ,
- What do you think about ... ?

29.2.2 Ways of making proposals

A research proposal is a **document** that presents a plan for a project.

document 書類

(1) **Propose + that specifies** who will carry out the proposed action.

specify 具体的に示す

Example:
- We proposed that it would be better for him to go the other way.

(2) **Propose + ing** does not specify who will do the proposed action, but is often used when the proposer himself or herself will perform the task.

Example:
- We propose going the other way.

Difference between *propose*, *recommend* and *suggest*:

The difference between *propose* and *recommend* as verbs is that *propose* means to offer a plan or course of action, whereas *recommend* means to **praise** that action. To *suggest* is to imply something without saying it clearly.

praise ほめる

Text: Quantum computers

Quantum computing is the use of the features of quantum states, such as **superposition** and **entanglement**, to perform **computation**. Quantum computers are machines that can perform quantum computing. The quantum in "quantum computing" refers to the quantum mechanics used by the system to compute outputs. A quantum is the smallest **discrete unit** of any physical **attribute** in physics. It usually refers to the properties of electrons, neutrinos, and photons, which are atomic or **subatomic** particles.

Information is **encoded** as bits in a normal computer, which are interpreted as 0 or 1. This information is stored in a quantum computer in a slightly different format: quantum bits, or "qubits." This qubit is physically capable of being in one state, another, or someplace in between. Quantum computers, therefore, in contrast to classical computers, **execute** calculations based on the probability of an object's state before it is measured, rather than only considering 1s or 0s, allowing them to process **exponentially** more data. On the other hand, classical computers use the definite position of a physical state to perform logical operations. Basic quantum computers that can execute specific computations have already been developed by scientists, but a real quantum computer is still years away.

Quantum computers could become well adapted to modelling complex systems such as economic market forces, astronomical dynamics, and **genetic mutation** patterns in living creatures, **to mention a few**. Although academics disagree on this topic, it is suggested that large-scale quantum computers will assist AI (artificial intelligence) **and vice versa**. Quantum computing thus opens up previously unimagined opportunities for addressing society's most difficult problems.

superposition 重ね合わせ
entanglement もつれ
computation 計算
discrete unit 離散単位
attribute [ˈætrɪbjuːt] 特性
subatomic 亜原子の
encoded 入力する（反意語は decode）

execute [ˈeksɪkjuːt] 実行する
exponentially 指数関数的に

genetic mutation 遺伝子変異
to mention a few これらは数多くある例の一部に過ぎない
and vice versa その逆も真. ⇒ⓖ16.9.9
address ... problems ⇒ⓖ16.5

Exercises

Understanding the text
Q1 What is the difference between a *bit* and a *qubit*?

Grammar point
Q2 Complete the sentences using the correct form of the words in the brackets.

　i. It has been proposed that **radiation pressure** may _____ (propel) a spaceship through the solar system.

　ii. Avogadro suggested that all gases _____ (occupy) the same volume under the same temperature and pressure conditions contain the same number of atoms or molecules.

　iii. The committee recommended _____ (select) these values.

Writing
Q3 Write a paragraph recommending the possible measures that a government may **implement** to further research into quantum computers.

> radiation pressure 放射圧
>
> implement 実施する
> research into … ～の研究だが，research of ではない
> ⇒ G 7.5

第29章　示唆

　何らかの示唆をする際，他の人に考えてほしいある企画やアイデアを示すことになります。今後の研究に関する示唆を論文や発表で示す必要があります。

29.1　示唆を表すことば

　決定を下す前に他の可能性を徹底的に検討する必要があります。提案書を見ると，最良の解決法を導く前に数多くの可能な案が検討されています。提案書は，調査とデータに基づいた最終的な推奨を行うため，より明確な説得力をもちます。

　何かをしなければならないと話し手が決定した場合に must を使います。それに対し，何かをしなければならないと話し手以外が決定した場合には have to を用います。

　例：
　・The electric field **must** be defined for all points in the region.（電場は領域内のすべてのポイントに対して定義しなければならない。）
　　・The particle itself does not **have to** rotate around the center.（粒子それ自体は中心の周りを回転する必要はない。）

　should は何かを示唆する，あるいは提唱する際に使います。論文またはレポートの推奨セクションで，should は受動態で使われます。

　例：
　・During the parachute landing, the legs **should be** together.（パラシュートで着地する際，両脚が離れているのは望ましくない。）
　・This is the maximum stress to which the steel wires **should be** subjected.（これは，鋼線に与える必要のある最大応力である。）

　示唆を表す動詞や重要な形容詞を，相手にしてほしい行動と組み合わせれば，示唆が強まります。命令するのではなく，アドバイスや勧めを強調するのです。

　例：
　・I **suggest** that you try improving the engine efficiency.（エンジン効率を上げるようにしてみてはどうでしょう。）
　・I **recommend** that you extract the energy as heat from the water.（水から熱エネルギーを取り出してはどうでしょう。）

29.2　示唆の仕方

　示唆にはいろいろな仕方があります。自分自身や同じグループの人たちへの呼びかけとしては，以下の語句が役立つかもしれません。

　・Why don't we do …?（…しましょう。）

・Let's try doing … （…をやってみましょう。）
・I would suggest that we do … （…してみてはどうでしょう。）
・Let's do … .
・What/How about doing … ?
・We could do … （…できるのでは。）

示唆を求める質問には次のようなものがあります。

・What should I (or should we) do?
・What do you think I should do?
・What do you suggest we do?

29.2.1　別の案を示す

他の案を示すときには，とても丁寧でかつ主張が強すぎない以下の表現が便利です。

・How about … ?
・（I think/I believe）it might help to …
・It might be useful to …
・I would suggest …
・It'd be better to …
・One thing you could do is work on … ,
・What do you think about … ?

29.2.2　提案の仕方

研究提案書はプロジェクトの計画を提示する文書です。

⑴　**propose + that** の表現は，提案を実行する人を明示します。

例：
・We **proposed that** it would be better for him to go the other way. （別の案の選択を彼に提案した。）

⑵　**propose + -ing** の表現では提案を実行する人を明らかにしませんが，多くは提案者自身が実行します。

例：
・We **propose going** the other way. （別の案の選択を提案する。）

propose と recommend, suggest の違い

propose と recommend の違いは，propose が計画や実行の案を提示することを意味するのに対し，recommend はその行動を認め，評価することを意味します。suggest は，明確にそれとは言わずに暗に示すことです。

テキスト：量子コンピュータ

　量子計算とは，重ね合わせやもつれといった量子状態の特徴を利用することです。量子コンピュータは量子計算のできる機械です。"量子計算"の量子とは，システムが出力を計算するのに使う量子力学のことをいいます。量子とは，物理学における物理属性の最小の離散単位であり，原子または亜原子の粒子である，電子やニュートリノ，光子の性質を指すのがふつうです。

　情報はふつうのコンピュータではビットとしてエンコードされ，0または1として解釈されます。この情報は，量子コンピュータでは，量子ビット（キュービット）という少し変わった形式で保存されます。このキュービットは，ある状態，別の状態，またはその間のどこかに物理的に存在できます。そのため，従来のコンピュータとは対照的に，量子コンピュータは，1または0だけを考えるのではなく，測定前の物体の状態の確率に基づいて計算を行い，多くのデータを指数関数的に処理できます。それに対し従来のコンピュータは，物理状態の定まった位置を使って論理演算を行います。特定の計算を行える簡単な量子コンピュータは科学者がすでに開発していますが，実用可能な量子コンピュータの実現にはまだ何年もかかります。

　量子コンピュータは，経済の市場原理や天体力学，生物の遺伝子変異パターンなど多くの複雑なシステムのモデル化にうまく適応する可能性が十分にあります。専門家の意見は一致していませんが，大型の量子コンピュータがAI（人工知能）をサポートする，AIも大型の量子コンピュータをサポートする，との見解が示されています。量子コンピュータは，社会の直面する最も困難な問題に取り組むという，これまで想像もできなかったことを可能にしてくれるかもしれないのです。

Linking words

Transition words and phrases, also known as linking or connecting words, are used to connect various ideas in a text. They aid the reader's comprehension of the arguments by expressing the relationships between different sentences or parts of sentences. Without linking words, the information presented is **merely** a collection of words. If you do not want your writing to be **awkward** or disconnected, make good use of linking words and expressions. For the sake of clarity, it is critical to understand how linking words can be used to combine ideas in writing - and thus ensure that ideas within sentences and paragraphs are elegantly connected.

transition word 転換語

merely … 〜にすぎない
awkward ぎこちない

30.1 Transition or linking words

In all writing, it is critical to show how ideas are connected. Transition or linking words are words that clarify these connections. Conjunctions are usually used to link clauses in the same sentence. To link two different sentences, non-conjunction words or phrases are also used.

(1) Words that show contrast

contrast 対比

although, but, despite, even though, however, in contrast, in spite of, nevertheless, on the contrary, on the other hand, whereas, yet

Examples:
- Although there is no such thing as a truly ideal gas in nature, at low enough **densities**, all real gases approach the ideal state.

density 濃度

- The magnitude of the two charges is the same, but the sign is opposite.
- If there is **interference**, there will be areas that are dark to the observer, even though they are illuminated.

interference 干渉

- In contrast to the previous hypothesis, the new scientific study has revealed unexpected findings that challenge the existing theories.
- In spite of these local variations, the average dipole field changes only slowly over such short time intervals.
- A **virtual image** exists only within the brain but nevertheless is said to exist at the perceived location.

virtual image 虚像

- However, both processes have the same initial and final states.
- Humans have been around for about 10^6 years, whereas the universe has **been around** for about 10^{10} years.

be … around 存在している

- On the other hand, the total annual electrical energy production in the United States corresponds to a matter mass of only a few hundred kilograms.
- His entire career was based on physics, yet he had no idea what physics was.

(2) Words that give additional information:

also, besides, furthermore, in the same way, likewise, moreover, similarly

Examples:
- The graph also reveals how fast the kangaroo moves.
- Furthermore, the Law of Conservation of Energy states that energy ca<u>nn</u>ot be created <u>or</u> destroyed, but only transformed from one form to another.

<u>not … or</u> ⇒ ⓖ 16.7

- In the same way, if a cup of hot coffee is left on the table, its temperature will drop until it reaches room temperature.
- Moreover, the **net** force on the test charge would be directed away from the sheet.

net 正味の

- *Similarly*, when the valve on a garden hose is opened fully, a wave travels in water along the hose.

(3) Words that show sequence in time:

sequence 連続

after, *afterwards*, *as soon as*, *now*, *once*, *then*, *when*, *while*

Examples:
- It is possible to accelerate *while* maintaining constant speed.
- Just *afterwards*, the current becomes zero.
- Charge begins to flow *as soon as* the circuit is completed.
- *Once* **inside**, the particle is **shielded** from electric fields.
- The **cyclotron** is *then* operated by varying the magnetic field.
- *Now* consider the current in the lower half of the wire.
- *When* the switch is turned off, the current in the **resistor** begins to increase.

once inside 中に入ると同時に（＝once it is inside）
shield 保護する
cyclotron 原子破壊機
resistor 抵抗器

30.2   Past participial phrase

To connect two short sentences the subject + **passive verb** can occasionally be replaced with the **past participle** alone, making the expression more compact.

passive verb 受け身動詞
past participle 過去分詞（pp）

Example:
- The **defense cannon**, located at **sea level**, **fires** <u>fast-moving</u> balls. (The defense cannon, which is located at sea level ...)

defense cannon 防衛大砲
sea level 海抜
fire 放つ
fast-moving ⇒ 16.1

The past participle can be used at the start of a sentence as well. The subject of the participle must be the same as the subject of the verb of the main clause.

Example:
- Located at sea level, the defense cannon fires balls with a high initial speed.

30.3   Present participial phrase

In a non-finite relative phrase, the **present participle** can appear if the **preceding** noun is the subject of the participle. This type of construction can also be used to link sentences and topics.

present participle 現在分詞
preced 先行する

Examples:
- This gives a result confirming the value found before.
- This gives a result which confirms the value found before.

30.4   Linking words and phrases in a presentation

There are some words and phrases that can be used to highlight to the audience when the speaker is moving to the next slide, topic, or figure.

(1) Moving to the next topic

- Let's move on to the next section of the talk ...
- We can now turn to the next part of the talk..

(2) Referring to a previous topic (slide)

- As I have been explaining ...
- As I explained in ...

(3) Summing up

- I think I've now covered the main points ...

Chapter 30

Text: Laser technology

Energy is released in the form of radiated light in a laser. Lasers were created for the first time in 1960 and have evolved into **versatile** tools in recent years, resulting in a true technological revolution.

Atoms release rays of varying lengths, preventing a concentrated beam of light from **forming**. The laser causes all of the atoms to move in the same direction. Because the range or area of light spreading out is relatively small, it can go very long distances.

Scientists have produced different types of lasers since its first discovery, including those that use **luminescent** crystals, **luminous** glass, a mixture of various gases, and finally **semiconductors**. At 1500 degrees Celsius, a CO_2 gas laser can cut through **brick** and rock. Semiconductor quantum generators, which were first invented in 1962 at the Lebedev Institute of Physics, are unique among **optical** generators. While the ruby crystal laser is tens of centimeters long and the gas generator is nearly a meter long, the semiconductor laser is only a few tens of millimeters long, with a radiation density hundreds of thousands of times that of the best ruby laser. However, the most **intriguing** feature of the semiconductor laser is its ability to convert electrical energy straight into light waves. The semiconductor laser, with an efficiency of about 100% compared to a maximum of around 1% for other types, opens up new possibilities for manufacturing extremely **cost-effective** light sources.

The beam of a laser can be concentrated very precisely. Because of its precision thus as a measurement tool, scientists have been able to calculate the speed of light more precisely than ever before, as well as determine the Moon's exact distance from the Earth, thanks to laser reflectors installed on the Moon by American astronauts. **Surgeons** using the laser as a **surgical** knife have discovered that it can make bloodless **incisions**, and it is proving to be quite useful in delicate eye **surgery**.

A laser system's conception, construction, and operation represent a significant technological achievement. However, the laser's greatest widespread application in the future will be in the **sphere** of communication. **Solar radiation** will be converted into laser beams by lasers **mount**ed on Earth **satellites**. Scientists predict that one day, a single laser beam will be used to convey millions of phone calls or a thousand television programs at the same time. It will be used to communicate quickly across continents, beneath the sea, between the Earth and spaceships, and between persons in space. The potential value of these applications continues to **drive** new laser technology advancements.

versatile 用途の広い

form 形成される

luminescent [ˌluːmɪˈnesnt] 発光性の
luminous 発光する
semiconductor 半導体
brick 煉瓦
optical 光の

intriguing 興味深い

cost-effective 費用対効果の高い⇒C16.1

surgeon 外科医
surgical 外科の
incision 切開
surgery 手術

sphere 球
solar radiation 太陽輻射
mount 搭載する
satellite 衛星

drive 推進する

Exercises

Understanding the text

Q1 What does laser stand for?   Write a paragraph on lasers using the following words in *italic*. Use as many linking words as possible to make the text more readable.

*coherent, concentrated, directional, in all directions, in one direction, laser, light **amplification**, organized, photon, stimulated emission of radiation, wavelength*

amplification 増幅

Grammar point

Q2 Join the sentences using transition words or past participles to form one complete sentence:

i. **Frictional forces** do not act on the objects in the system. The system is isolated from its environment.

frictional force 摩擦力

ii. Project Seafarer was an ambitious program to construct an enormous antenna. It was buried underground.

第 30 章　連結語

　転換語（句）（transition word）は，連結語（linking word）とも呼ばれ，テキスト内のさまざまな
アイデアを関連づけます。これらの語は異なる文や文の一部の関係を表現し，読者が論の展開を
理解するのを助けます。連結語がなければ，提示された情報は単なる語の集積にすぎません。文
章をぎこちなくぶつぶつ切れたものにしないためには，連結語や表現をうまく活用しましょう。
わかりやすい文章にするには，どのように連結語を用いれば文章中のアイデアを組み合わせるこ
とができるのか，言い換えると，文章やパラグラフ内のアイデアを適切につなげることができる
のかを理解する必要があります。

30.1　転換語または連結語

　どのような文章を書く場合でも，アイデアとアイデアのつながりを示すことが求められます。
このようなつながりを明確にする言葉を転換語や連結語と呼びます。接続詞は通常，同じ文の節
をつなぐために使われます。異なる 2 つの文をつなぐには，接続詞以外の語（句）も使われます。

(1)　対比を示す語

although, but, despite, even though, however, in contrast, in spite of, nevertheless, on the contrary, on the
other hand, whereas, yet

例：

- **Although** there is no such thing as a truly ideal gas in nature, at low enough densities, all real gases
  approach the ideal state. (真の理想気体は自然界に存在しないが，濃度を十分に低くすれ
  ば，実在気体をすべて理想気体に近づけることはできる。)
- The magnitude of the two charges is the same, **but** the sign is opposite. (2 点電荷は大きさは同
  じだが，符号が反対となる。)
- If there is interference, there will be areas that are dark to the observer, **even though** they are illuminated. (干渉が起これば，光っていても観察者には見えない暗い場所が生じる。)
- **In contrast to** the previous hypothesis, the new scientific study has revealed unexpected findings
  that challenge the existing theories. (これまでの仮説とは対照的に，新しい科学的研究によ
  り明らかになった，予想していなかった結果は，既存の理論に疑問を投げかける。)
- **In spite of** these local variations, the average dipole field changes only slowly over such short time
  intervals. (個々には違いが生じるが，比較的短い時間の中での双極子場の変化は，平均し
  て見ればとてもゆるやかだ。)
- A virtual image exists only within the brain but **nevertheless** is said to exist at the perceived location. (虚像は脳内に存在するにすぎないが，知覚している人は，それが知覚している場所
  に存在していると考えているようだ。)
- **However**, both processes have the same initial and final states. (しかし，異なる 2 つのプロセス
  は，初期状態と最終状態で変わらない。)
- Humans have existed for about 10^6 years, **whereas** the universe is about 10^{10} years old. (人類は約
  10^6 生きてきた。一方宇宙が誕生して約 10^{10} 年になる。)
- **On the other hand**, the total annual electrical energy production in the United States corresponds
  to a matter mass of only a few hundred kilograms (これに対し，アメリカ合衆国の年間の電気
  エネルギー生産は，わずか数百キログラムの物質の質量に相当するにすぎない。)
- His entire career was based on physics, **yet** he had no idea what physics was. (彼は物理学をずっ
  と学び，研究してきたが，にもかかわらず，彼には物理学とは何かがわかっていなかっ
  た。)

(2)　追加情報を与える語

also, besides, furthermore, in the same way, likewise, moreover, similarly

例：

- The graph **also** reveals how fast the kangaroo moves. (このグラフはまた，カンガルーの移動
  の速さを示している。)
- **Furthermore**, the Law of Conservation of Energy states that energy cannot be created or destroyed,
  but only transformed from one form to another. (さらに，エネルギー保存の法則は次のように
  述べる「エネルギーを生み出すことも破壊することもできない。エネルギーは一つの形
  体から別の形体に形を変えるだけだ」と述べられる。)
- **In the same way**, the temperature of a cup of hot coffee, left sitting on the table, will fall until it
  also reaches room temperature. (同様に，テーブルの上のホットコーヒーの温度は，室温に
  達するまで下がっていく。)

- **Moreover**, the net force on the test charge would point away from the sheet. （さらに，試験電荷にかかる正味の力は，シートから離れた方向を指す。）
- **Similarly**, when the valve on a garden hose is opened fully, a wave travels in water along the hose. （同様に，ガーデンホースのバルブが十分に開くと，圧力波はホースの水の中を伝わる。）

⑶ 時間の連続を示す語

after, afterwards, as soon as, now, once, then, when, while

例：

- It is possible to be accelerating **while** traveling at constant speed. （一定の速度で動きながら加速度をつけるのは可能だ。）
- Just **afterwards**, the current becomes zero. （その直後，電流はゼロになる。）
- Charge begins to flow as **soon as** the circuit is completed. （回路が完成するとすぐに，電荷が流れ始める。）
- **Once** inside, the particle is shielded from electric fields. （中に入ると，粒子は電界から保護される。）
- We **then** operate the cyclotron by varying the magnetic field. （次に，磁場を変化させて原子破壊機を操作する。）
- **Now** consider the current in the lower half of the wire. （今度は，ワイヤーの下半分の電流について考える。）
- **When** the switch is closed, the current in the resistor starts to rise. （スイッチをオフにすると，抵抗器の電流が上昇を始める。）

30.2　過去分詞を用いた表現

短い 2 文を連結する際，「主語 ＋ be ＋ 過去分詞（受身の表現）」を過去分詞だけで置き換え，より簡潔な文を作ることがあります。

例：

- The defense cannon, **located** at sea level, fires fast-moving balls. （ = The defense cannon, which is **located** at sea level ...）（海抜の高さに設置された防衛大砲は，高速のボールを発射する。）

過去分詞は文頭でも用いられます。その主語は主節の動詞の主語と一致していなければなりません。

例：

- **Located** at sea level, the defense cannon fires balls with a high initial speed. （海抜の高さに設置された防衛大砲の発射初速度は高い。）

30.3　現在分詞を用いた表現

時制（＝現在，過去のような時間表現）をもたない関係詞句では，先行する名詞が分詞の主語であれば現在分詞が現れることがあります。この種の構造は文とトピックを結びつけるためにも用いることができます。

例：

- This gives a result **confirming** the value found before. （これにより得られる結果は，以前見出した値を確定する。）
- This gives a result which **confirms** the value found before.

30.4　プレゼンテーションの際のつなぎ表現

プレゼンテーションの際，次のスライドやトピック，図に移行する時のつなぎの語句により聴き手の注意を引きつけます。

⑴ 次のトピックへの移行

- Let's move on to the next section of the talk ...
- We can now tunr to the next part of the talk ...

⑵ 前のトピック（スライド）への言及

- As I have been explaining ...
- As I explained in ...

(3)　要約

・I think I've now covered the main points ...

テキスト：レーザー技術

　エネルギーは放射光としてレーザー内に放出されます。1960年に初めて誕生したレーザーは最近用途が拡大し，技術革命に貢献しています。

　原子は波長の異なる光線を放出し，集束光線もしくは集束ビームの形成を防いでいます。レーザーのはたらきにより原子はすべて同じ方向に移動します。光が拡散する範囲は比較的狭いので，光はかなり遠くまで達することができます。

　レーザーの発見以来，科学者はあらゆるタイプのレーザーを作ってきました。たとえば，発光結晶や光ガラス，さまざまなガスの混成，それから半導体です。CO_2 ガスレーザーは，摂氏1500度で，煉瓦や岩を通過することができます。1962年に Lebedev 物理研究所で作られた半導体量子生成器は，光生成器の中で独特の特徴をもっています。ルビー結晶レーザーが何十センチもの長さを，ガス生成器が約1mの長さをもつのに対し，半導体レーザーはほんの数ミリメートルの長さでしかないのに，どんな精巧なルビーレーザーよりも数十万倍の放射線密度をもっています。しかし，半導体レーザーの最も興味深い性質は，電気エネルギーを光の波に転換できることです。他のレーザーがせいぜい1%の効率しかもたないのに対し，半導体レーザーのもつ効率はほぼ100%であり，非常に費用対効果のすぐれた光源を作り出す可能性があります。

　レーザービームは非常に正確に集中させることができます。その精密さゆえに，科学者たちは計測ツールとして，これまで以上に正確に，光の速さを計算できるようになりました。また，アメリカの宇宙飛行士が月に設置したレーザー反射器のおかげで，地球と月との距離をより正確に測ることができるようになりました。外科医はレーザーをナイフ替わりに使って，血を出さずに切開手術を施しています。細心の注意が必要な目の手術にも役立てられようとしています。

　レーザーについての考え方やレーザーの製作，その操作は，人類におけるたいへん重要な技術の進歩を示しています。しかし，レーザーの最も重要な応用はコミュニケーションの分野においてでしょう。地球衛星に搭載したレーザーにより，太陽輻射をレーザー光線に転換できるのです。科学者の予測では，将来，レーザービーム1個で数百万回の通話や1000個のテレビ番組を同時に発信できるようになります。それにより大陸間で，また海中で，地球と宇宙船とで，宇宙にいる人同士での交信が可能となるのです。これらの可能性を秘めた応用技術の価値の大きさが，新しいレーザー技術の促進に拍車をかけ続けています。

付録G

文 法 Grammar

　文法の学習は技術英語論文作成にあたり重要です。例文末尾（　）の章・節・項番号は，本文と対応しています。

G1　定冠詞と不定冠詞

　冠詞は，それを伴う名詞が特定のものを指すかどうかを示します。英語には定冠詞と不定冠詞の2つの冠詞があります。

G1.1　定冠詞

　定冠詞（the）は，名詞がすでに言及されているか，言及されていなくても読者あるいは聴き手によく知られていることを示します。単数の可算・不可算名詞と複数の可算名詞の前に the は現れます。

　強く認識してほしいのは，受け手にとって特定である場合に the が用いられることです。日本人は the を使いすぎる傾向がありますので，「この名詞は特定だろうか」と自問したうえで使ってください。

　例 i ：
　・I read **the** book yesterday.

　この場合，読者（聴き手）は，どの本を指しているのかを知っています。

　例 ii ：
　・**The** two of them attended the party.
　　（パーティーに出席したのは彼ら2人）
　・Two of them attended **the** party.
　　（パーティーに出席したのは彼らのなかの2人）

前置詞句により修飾される名詞は the をとることが多いです。

　例：
　・**the** return of the birds
　・**the** beauty of bird song

G1.2　不定冠詞

　不定冠詞（a と an）は，名詞が読者（聴き手）に特定されていない場合に用いられます。

　例：
　・I have given a/**the** solution to the problem.
　　⇒ a＝複数ある解決策の中のある1つ　　the＝1個しかない解決策

G2　可算・不可算名詞

　名詞には可算名詞と不可算名詞の2種類があり，一定の形をもつものを可算名詞，もたないものを不可算名詞と呼びます。次の例で違いを確認します。

　例：
　・He ate **a** chicken.（ひとりで鶏を一羽食べた！！）
　・He ate **chicken**.（鶏の肉を食べた。）

　a chicken は「鶏一羽」を指します。1羽，2羽と数えられますので可算名詞です。一方，「鶏の肉」の形は，一定でなく，1個，2個と数えられませんので不可算名詞です。鶏肉2切れは "two slices of chicken" と表現します。

G2.1　不可算名詞の例

液体や気体，抽象名詞（美，自由など），学問名（数学，物理学など），自然現象（重力，地震など）などは一定の形をもたず，不可算名詞として扱われます。

情報（information）は抽象名詞であり，不可算名詞です。3つの情報は "three pieces of information" と表現されます。

G2.2　可算名詞の用法

可算名詞には大きく3つの用法があります。

(1)　特定ではない名詞に言及

　　例：
　　・Energy produced at a power plant can be used to run a home computer.（発電所で生成されたエネルギーを使い，家庭用コンピューターを動かすことができる。）

(2)　初めての可算名詞への言及
　　2度目からは the を伴います。

　　例：
　　・They bought a new machine. In comparison to the previous one, the new one was more efficient.（彼らは新しい機械を購入した。以前と比較して新しいものは効率性が上回っていた。）

(3)　〜につき

　　例：
　　・four thousand yen a kilogram
　　・100 km an hour
　　・twice a month

G2.3　単数にも複数にもなる集合名詞・群名詞

家族やチーム，クラス，家具などの集合名詞・群名詞（group nouns）は，単数としても複数としてもとらえることができます。次の例で考えましょう。

　　例：
　　・This team is the best.
　　・This team are all fast runners.

チームを一つのチームとみる（単数扱い）か，チームの個々のメンバーの総体としてみる（複数扱い）かの違いです。同様に，「十年の月日」の「十年」も単数と複数のいずれでもとらえることができます。

　　例：
　　・Ten years is a long time.（十年一昔）
　　・Ten years have passed since then.（あれから十年の日々が経った。）

G2.4　a dog, the dog, dogs, the dogs

「犬は人間の友」ということばがあります。英語ではどのようにいうのでしょうか？

　　i ．A dog is a friendly animal.
　　ii ．Dogs are friendly animals.
　　iii ．The dog is a friendly animal.
　　iv ．The dogs are friendly animals.

iv．以外はいずれも使用可能です（iv．は特定の複数の犬に言及します）。一般には ii．が多く用いられます。i．も可能ですが，「ある犬は人間の友」とも解釈できてしまいます。論文では iii．が使われます。ただし，論文以外では，iii．は特定の犬に言及しているとも解釈できますので，一般には ii．がよく用いられることになります。

　　v ．I like a book.

vi.　I like books.

「私は本が好きです」といいたければ，vi. の表現にしましょう。v. は「私には好きな本が（1 冊）あります」と受け取られる可能性が高く，「その本のタイトルは？」と聞かれるかもしれません。

G2.5　その他有益な知識

G2.5.1　[a ＋ 形容詞 ＋ 不可算名詞]

例：
・She was welcomed with warmth.
・She was welcomed with a surprising warmth.
　（さまざまな温かさのなかの一つとしての a surprising warmth）

G2.5.2　[a 名詞 ＋ a 名詞] ではなく [a 名詞 ＋ 名詞]

例：
・Format of a report and paragraph（1 title）
・When analyzing information in a line or bar graph, the time periods should be noted as well as the amount increased or decreased during that interval.（7.2）
・the old and new worlds

G3　代名詞

G3.1　you, we, one

学術論文で特定の個人に言及しない場合には，you, we, one を主語として用いることができます。one はあらたまった表現で，あまり使われません。一般には you がよく用いられます。書き手（話し手）も含まれていることをはっきりさせる場合には we が適切です。
　理工系の論文では you は用いず，we を使います。

例：
・You're healthier as you undergo moderate exercise.（適度な運動もすればより健康になる。）
・We grow mango, a tropical fruit, in Amami Oshima.

書き手（話し手）が含まれない場合には they を用います。

例：
・They play cricket in India.

　理工系の学術論文では you は使いませんが，本書は論文ではなく教科書（テキスト）ですので，you の使用は問題ありません。

例：
・The conclusion should contain whatever you want the reader to remember.
・To understand how to write a report, we need to first understand how to write a paragraph showing its significance to the report as a whole and then how to link it to the next section/paragraph.（1）
・When writing the recommendation part, one needs to state which other actions should be taken as a result of the research.（9）
・Recognizing one's social responsibilities as a scientist is an important first step.（17 Text）

G3.2　one と ones

可算名詞でどの名詞を指すのか明らかなとき，その名詞を one, ones で置き換えることができます。

例：one = section
・It is commonly thought that the Discussion section is the most challenging one to write.（6.1）

例：ones = hypotheses
・As astronomer Carl Sagan wrote: "Science invites us to let the facts in, even when they don't conform to our preconceptions. It counsels us to carry alternative hypotheses in our heads to see which ones best match the facts." (16 Text)

G3.3 any/some/every/no ＋ body/one/thing/where

たとえば somebody（誰か），something（何か），somewhere（どこか）のようになります。

例：
・I saw somebody outside the room.
・I think that I need something more.
・I've got to go somewhere to meet her.

G3.4 疑問文中の some

疑問文中の some は「〜を勧める」を意味します。

例：
・Would you like any more coffee?（お飲みになりますか。）
 ⇒「飲むのか飲まないのか」を問う普通の疑問文

・Would you like some more coffee?（お飲みになりませんか。）
 ⇒飲むことを勧めている

G3.5 否定文中の some

次の例を通して，否定文中の some の用法を確認してください。

(1) any の場合

例：
・I haven't found any books.
 ＝I have found no books.（本は一冊も見つからなかった。）という否定文になる

(2) some の場合

・I haven't found some books.
 ＝There are some books I haven't found.（見つからなかった本があった。）という肯定の意味となる

G3.6 肯定文中の［any ＋ 単数形］（複数形）

肯定文中の［any ＋ 単数形（複数形）］は（対象の中の）どれでもよいことを示します。

例：
・"Which book on the shelf do you want to read?" "Any (book) will do.（どれでもいいよ。）"
・Rehearse to ensure that the timing of the presentation is accurate and that there is also enough time to make any necessary modifications. (13.5)

G3.7 a few, few /a little, little

a few/a little は「少ない」という否定的な意味をもちません。十分ではないが「いくらかはある」という肯定的な意味をもちます。「ほとんどない」という否定的な意味をもつのは few/little です。

例：
・Good readers begin by previewing the text, i.e., looking over the entire passage for a few moments.（しばらく）(2.1)
・There are few (if any) free electrons in an insulator.（絶縁体には（あるとしても）自由電子はほとんどない。）

G3.8 every, each, any

every と each は可算名詞の単数形と共に用いられ，「すべての〜」と，指す対象は同じです。ただし，each は個々の一人ひとりに焦点が当てられるのに対し，every は全体に目が向けられます。

例：それぞれの学生がもっている
・Each student has an English-English dictionary.

例：全員がもっている
・Every student has an English-English dictionary.

肯定文で用いられる any（単数・複数どちらの名詞もとれる）は，特定の名詞に言及しません。

例：
・Check to see that every sentence is related to the topic sentence.
・The whole report is divided into several sections, and each section is divided into paragraphs. (1)
・The paragraph may be organized according to any one of the following concepts: (1.1.2)
・The introduction should summarize the relevant literature and explain any important technical words which will be used extensively throughout the paper. (3.1)

G3.9 名詞に先行する代名詞

たとえば，He thought that Bacon would succeed. という文で，代名詞（he）が後ろの名詞（Bacon）を指すことはできませんが，次の文で his は Bacon を指します。

As an instance of his method, Bacon describes an experiment into the nature and cause of the rainbow. (27.1.2)

次の文と同じ意味です。副詞句が先行するこの種の文では，普通次のようには言いません。

As an instance of Bacon's method, he describes an experiment into the nature and cause of the rainbow.

G4 形容詞

形容詞は，名詞を修飾するか，動詞の後に用いられます。

(1) 名詞を修飾

例：
・a happy prince

(2) 動詞の後

例：
・The prince is happy.

上記のうちどちらかの用法しかない形容詞もあります。

(3) 名詞を修飾するが，動詞の後には来ない

例：
・a main street in London

The street is main in London. とはいえません。

(4) 動詞の後に来るが，名詞を修飾しない

例：
・He fell asleep.

an asleep boy とはいえず，a sleeping boy と表現します。

G5　副詞

副詞は，動詞や副詞，形容詞を修飾します。

(1)　動詞を修飾

　例：
　・He comes late.

(2)　副詞を修飾

　例：
　・She works very hard.

(3)　形容詞を修飾

　例：
　・You are very tired.

副詞のタイプ	例
どのように	She works hard.
程度・方向	The shaft was angled slightly upward.（シャフトはわずかに上向きに角度が付けられていた。）
頻度	He often comes late.
場所	We live here.
時	I saw John yesterday.

G5.1　副詞の位置

　副詞の取り得る位置は一様ではないため，学習者を混乱させることがあります。修飾する語（動詞・副詞・形容詞）のできるだけ近くに置くのが原則です。

G5.1.1　基本の位置

(1)　動詞の前

　例：
　・You must always come early.

(2)　文頭

　例：
　・Today I have a computer lesson.

(3)　動詞の後ろで文末

　例：
　・The water is rising fast.

G5.1.2　強調するときの位置

　場所と時間を表す副詞は文末に置かれることが多いのですが，強調表現では文頭に回すことができます。

例：
- **Here** I saw John yesterday.
- **Yesterday** I saw John here.

G5.1.3 副詞の順序

一文の中に複数の副詞があるときは通常，どのように（manner）– 場所（place）– 時（time）の順になります。

例：
- We work **hard**（manner）**in the laboratory**（place）**in the afternoon**（time）.

動詞が動きを表す場合には，場所 – どのように – 時の順をとります。

例：
- We run **in the field energetically every morning**.

run in/run to のように，走るという意味の動詞に対して「どこを／どこに」の情報が重要であり，したがって動詞の直後に置かれるのです。

複数の副詞の順序が変わるのは，その中の一つの副詞を強調する場合です。

例：
- **In the afternoon**, we work hard in the laboratory.（午後は，研究室で仕事をします。）
この文は，「午前中は別の場所で仕事をしているかもしれない」ということを暗示します。

G5.2 文副詞

動詞や副詞，形容詞を修飾するのではなく，副詞が文を修飾することがあり，文副詞と呼ばれます。

例：
- He didn't die **happily**.（最後の日々は幸せではなかった。）
  ⇒ happily は die を修飾している

- **Happily**, he didn't die.（幸いにも命を取り留めた。）
  ⇒ Happily は文全体を修飾している

G5.3 文頭での文副詞としての用法

例：
- **Interestingly**, three factors–the title, abstract and keywords–may hold the key to publication success.（10.3.3）
  = It is interesting that three …

- **Admittedly**, this is quite wrong.（Table 6.2）
  = It can be admitted that this is …

G5.4 高校時代には学ばなかったであろう副詞

例：
- The paragraph should contain a topic sentence, **preferably** at the beginning, which discusses the main hypothesis presented in the paragraph.（1.1.1）
- Readers need to see the meaningful combinations of words and phrases to **eventually** comprehend the passage completely.（2.2）
- A purpose is usually expressed by an adverbial clause which **typically** begins with the conjunctive phrases *so that* and *in order to*.（3.2）
- These days, you may use *may* and *might* **interchangeably**; however, there is a slight distinction between the two.（3.2box）
- It is better to save the readers from exerting any mental effort and at the same time influencing them to look **favorably** on the interpretation of the experimental data that is consistent with your own deductions.（7.4）

- Articles start with a summary paragraph called an abstract, ideally not more than 200 words. (10)
- So basically, it is the abstract which helps the reader determine whether they should read the entire article or not. (10)
- Physicists like to study seemingly unrelated phenomena. (11.2.3)
- By using concise and lucid language to confirm the necessity for a particular research goal, the researcher will be more likely to produce a high-quality technical document and thus promote more positively a future research course. (16)
- When a young scientist is trying to achieve success in his/her research, the preliminary setting of realistic goals is crucial, that is, essential or vitally important for the accomplishment of his/her tasks.
- Particularly in the Discussion section, it is essential that you purposively provide the readers with opportunities for improvement considering the limitations of your own research, especially with regards to methodology and in the interpretation of the results. (22.2)
- The abacus, slide rule, and, most likely, the astrolabe and Antikythera mechanism (circa 150-100 BC) are examples of early mechanical computing machines. (25 Text)
- For example, if the charge is uniformly distributed over a sphere, we can enclose the sphere in a spherical Gaussian surface. (27.1.1)
- Virtual reality in education is on the horizon, and it will undoubtedly alter the way we live. (28Text)

G5.5　especially の用法

「私は相撲が好きです。特に宮城野親方（元白鵬）が」という文を "I love sumo wrestling. Especially (I love) the former Hakuho." と英文にするでしょう。しかし，especially を文頭で用いることはできません。

例：
- Keywords are very important elementss of an abstract; they help researchers discover content, especially through a search engine. (10.3.3)

文頭で使うには particularly か in particular を使いましょう。

例：
- Write a paragraph discussing a computer operating system. In particular, note the problems or features that need to be improved for such a structure. Present one or two ideas that can solve such problems. (Exercises 16)

G5.6　accordingly（それに応じて）

以下は，高校時代には見たことのない表現かもしれません。

例：
- In order to be productive, they therefore need to plan their work and execute it accordingly. (20 Text)
- It started raining hard; accordingly, the game was canceled.

G5.7　比較級・最上級

不規則な変化をする形容詞があります。

原級 （positive）	比較級 （comparative）	最上級 （superlative）
little	less	least
far	farther/further	farthest/furthest

(1)　数えられる名詞のとき
　　⇒可算名詞の前におく

例：
　　・You need fewer books.

⑵　数えられない名詞のとき
　　⇒不可算名詞の前におく

例：
　　・You need less water for this plant.

⑶　より遠く，さらに
　　⇒空間の表現に用いる

例：
　　・He went farther down the road.

⑷　質的により遠く（深く）
　　⇒質や程度などの表現に用いる

例：
　　・I wanted to discuss it further.

G5.7.1　最上級と of / in
⑴　of

例：
　　・Of the three samples, this is the most interesting.
　　　「複数の A の中で B が一番〜」という構文で用いられます。

⑵　in

例：
　　・Who is the most famous scientist in your country?
　　　「A（場所，グループなど）に所属する中で B が一番〜」という構文で用いられます。

G5.7.2　much/ much more/ far/ a bit/ a lot/ a little/any ＋ 比較級

例：
　　・The drift speed is much smaller than the effective speed.（〜よるはるかに）
　　・She speaks German far better than I do.
　　　= She speaks German a lot better than I do.

a lot は much や far より口語的な表現です。

G5.7.3　by far ＋ the ＋ 最上級
by far は「断然」の意味をもちます。

例：
　　・John is by far the best singer.（ジョンは飛び抜けて歌がうまい。）

G5.7.4　still/even ＋ 比較級
still と even は「さらに」の意味をもちます。

例：
　　・John is tall, and Mike is still taller.
　　　（さらに，一層）
　　・It could get even worse and spark a major confrontation.（事情はさらに悪化し，ゆゆ
　　　しい対立を引き起こす可能性がある。）

G6　接続詞
　　重文と複文に用いられる接続詞については，セクション 4.2 と 24.1 で扱いました。また

G15.1 でも取り上げます。ここでは，接続詞の用法を 2 つ紹介します。

G6.1.1 because と as, since の違い

because は文の中心が理由である場合に用いられるのに対し，as と since を用いる文では，理由は文の中心ではなく既知の内容となっています。

例：
- He had to do his assignment right away because it was too late. （彼がすぐに任務を遂行しなければならなかったのは，手遅れだったからだ。）
- She was promoted above four other candidates as she was the most familiar with the system. （システムに最も精通していたため，彼女は他の 4 人の候補者よりも昇進した。）
- Since she knew the system best, she was promoted over the four other candidates. （システムを最もよく知っていたので，彼女は他の 4 人の候補者よりも昇進した。）

G6.1.2 and that の用法

動詞が 2 つの節を目的語としてとるとき，2 つ目の節の前には必ず that を置かなければなりません。でないと，2 つ目の節は動詞とは関係のない独立した文だと誤って解釈される可能性があります。

例：
- Rehearse to ensure that the timing of the presentation is accurate and that there is also enough time to make any necessary modifications. (13.5)

この文で and that の that を省くと，there is … が ensure の目的節であることが保証されません。

G6.1.3 文頭の and, but, or

接続詞の and, but, or を論文の文頭に用いることはできません。

G7 前置詞

前置詞は名詞や名詞句の前に置かれます。

例：
- Do in Rome as the Romans do. （郷に入っては郷に従え。）
- I'm against your going alone. （あなたが一人で行くのには反対です。）

7.1 前置詞の後に置けないもの

前置詞を動詞や節の前に置くことはできません。
- × I insist on go alone.
- ○ I insist on going alone.

- × We insist on he go alone.
- ○ We insist on him (or his) going alone.
  （his は「かたい」表現）

- × We insist on that he go alone.
- ○ We insist that he go alone.

例外として in that （～という点において）構文があります。

例：
- This equation is equivalent to the other equation in that either can be considered as the defining equation. (Table 6.1)

G7.2 with ＋ 名詞句 ＋ 分詞

以下の例は付帯状況の構文です。

例：
- I went out with the water of the bathtub running. （浴槽の水を出しっ放しで出かけ

た。）
・It will float in water with 90% of its volume submerged.

G7.3　from A to B to C

「東京から名古屋へ，そして京都へ行った」を英語では "I went from Tokyo to Nagoya to Kyoto." と表現します。以下の表現も同様です。

例：
・Moving from the past to the current situation to possible future outcomes.（1.1.2）

G7.4　to both A and B,　both to A and to B

「A に対しても B に対しても」を英語では 2 通りに表現できます。

例：
・*Can, must, have to,* and *need to* are common words that apply to both the present and the future（28.1）.

次のようにしても使えます。

例：
・*Can, must, have to,* and *need to* are common words that apply both to the present and to the future.

G7.5　of の過剰使用

X の Y を英語で表現する際，日本人は of を使いすぎる傾向があります。しかし，たとえば「ベートーヴェンのピアノソナタ」は英語で "a piano sonata by Beethoven" です。以下の文では of ではなく for を使います。

例：
・Science starts with problems. Conclusions are inferred from a tentative hypothesis, which is proposed and then tested to be able to provide an explanation for the facts.（18.1）（～の説明）
・Physical science is now challenged by the need for its use in technological applications.（18 Text）（～の必要）

G7.6　一点集中の on

on は「一点への集中」の意味でも使われます。concentrate on の on です。

例：
・In this presentation, I want to focus on ...（14.2（1））

たとえば「重力論」というタイトルの論文は重力についてのみ論じますので "On Gravity" と on を用います。

例：
・I will give a brief lecture on ...（14.2（1））
・Prepare a short presentation of around five minutes on an environmental problem.（13 Exercises）

G7.7　in turn

"as a result of something"（～の結果として，～を受けて）の意味をもつ "in turn" の例を紹介します。

例：
・Moving from the past to the current situation to possible future outcomes. Each part in turn（そして，次に，今度は）is the cause of the subsequent part.（1.1.2）

G8　関係詞（＝関係代名詞・関係副詞）

関係代名詞・関係副詞は本来日本語の文法にはなく，明治時代には，たとえば "a flower which he loved" を「彼が愛した所の花」と訳していました。英語の論文を書く際，関係詞は頻繁に用いられます。以下，関係詞を用いた重要表現をいくつか紹介します。

G8.1 前置詞＋関係詞

例：
・The question-and-answer session is a good place from where the conversation can start. (13.6)

G8.2 This is where ＝ This is the place where

例：
・The body of your paper is where you display your research, make pertinent points if you have an argument to make, contradict other arguments, and list your references. (14.1 (3))

G8.3 関係詞を用いた複雑な構文

たとえば以下の文は，次の2文を経てできました。

Make a list of at least three questions you expect to be asked at the conclusion of your presentation. (13.6)

・Make a list of at least three questions.
・You expect the three questions to be asked at the conclusion of your presentation.

この2文を関係詞でつなぐと次の文ができます。

Make a list of at least three questions which you expect to be asked at the conclusion of your presentation.

目的格の which を省略すれば最初の文となります。

次の文はどうでしょう。

Often, it is stated in the conclusion which hypotheses the writer believes have the strongest evidence.

これは，次の2文を関係代名詞（ここでは省略されています）で結んだ文です。

・Often, it is stated in the conclusion which hypotheses have the strongest evidence.
・The writer believes the hypotheses which have the strongest evidence.

このとき，which は疑問詞です。理解するのに時間がかかると思いますが，このような文をいつか書けるといいですね。

G8.4 that which ＝ what

that which は what におきかえられます。

例：
・The x-axis represents the independent variable, or that which/what can be changed. (7.2)

G9 時制

時制とは現実の時間を述べる表現のことで，過去・現在・未来を指します。

G9.1 現在時制

現在時制は現在形，現在進行形，現在完了形，現在完了進行形からなります。日頃起きている（行われている）ことは現在形で表します。事実や長く続いていること，自然の法則や真理，日常の行為や習慣なども現在形を用いて表現します。実際には未来に行われるスケジュールや時刻表も現在形が受け持ちます。

例：
　・The earth goes around the sun.
　・The plane flies for London tomorrow.

G9.1.1　現在進行形

現在進行形は発話時点での行為や確実に行う未来の計画を表現します。

例：
　・We are discussing the fundamental principles of physics.
　・We're leaving for London tomorrow.

G9.1.2　現在完了形

現在完了形は次の4つの状況を表します。

(1)　結果

例：
　・The winter has gone.（= We're now in the spring.）

(2)　完了

例：
　・I've finished writing a letter.

(3)　経験

例：
　・I've been to London twice.

(4)　継続

例：
　・I've lived in Tokyo for twenty years.

上述の例文はそれぞれ以下のように解釈され，これらはすべて現在とつながった表現です。

　・結果：今は春。
　・完了：今しがた書き終えた。
　・経験：現時点でのロンドンへの旅行回数は2回。
　・継続：私は今も東京に住んでいる。

次の2文の違いに注意しましょう。

　・I've lived in Tokyo for thirty years.
　・I lived in Tokyo for thirty years.

前者で私は今も東京に住んでいるのに対し，後者では私は今東京に住んでいません。

G9.1.3　現在完了形のもう一つの用法

現在完了形は，ある行為が終了したと想定された事態においても用いられます。

例：
　・Determine how much material you can offer in the time allotted after you have defined the goal of your presentation.（14.2 (2)）

次の文のように現在形を使ってもよいのですが，完了形にすることにより完了の意味が強く押し出されます。

例：
　・Determine how much material you can offer in the time allotted after you define the goal of your presentation.

G9.1.4　現在完了進行形

現在完了進行形は，過去のある時点で始まり今なお継続している行為を表現します。

例：

・It **has been raining** for three consecutive days.（3日間連続）

G9.2　過去時制

過去時制を用いる際にはいくつか考慮すべきことがあります。

G9.2.1　過去完了進行形

過去完了進行形は，過去のある時点まで続いていた行為を強調するために用いられます。

例：

・The people **had been travelling** for a while before a proper road was constructed.（き
ちんとした道路が建設される前のしばらくの間，人々は旅行していました。）

G9.2.2　ago と before の違い

例：

・He died three days **ago**.
　⇒現在から見た3日前

・He had died three days **before**（his son came to see him）.
　⇒過去のある時点から見た3日前

G9.2.3　「過去の過去」としての過去完了

過去完了時制は，過去のある時点で終了していた事態を述べるのに用いられます。過去
完了は「過去の過去」と捉えることができます。

例：

・Charles Babbage invented and built the first fully programmable mechanical computer,
dubbed "The Analytical Engine," in 1837. Several technologies that would later be useful in the development of practical computers **had emerged** by the end of the nineteenth century, including the punched card, Boolean algebra, vacuum tube（thermionic
valve），and teleprinter.（25 Text）

G9.3　未来時制

未来とは「現在からみた未来」ですが，「過去からみた未来」，あるいは，「過去におけ
る未来」もあります。「過去における未来」とは，過去のある時点で思い描かれた未来の
事態です。

例：

・In my junior high school days I thought I **would** be a doctor.
・Several technologies that **would** later be useful in the development of practical computers had emerged by the end of the nineteenth century, including the punched card, Boolean algebra, vacuum tube（thermionic valve），and teleprinter.（25 Text）

過去における未来は would で表現されます。

G10　受動態

技術英語では「we を主語とするのは避け受動態で表現するべき」との見解に対し，賛
否両論があるようです。この問題を扱った論文に樋口（1996）があります。ここでは，そ
の問題には触れず，一般に受動態が選択される状況を紹介します。

G10.1　行為者に言及しない受動態

行為者への言及が必要ない，行為者が不明または文脈上明らかな場合には，受動態が選
択されます。

例：
- The water was poured into the cup.
- The letter was delivered at 9 a.m.
- The structural beams of the building had been damaged.（建物の構造梁が損傷していた。）
- John is said to be a businessman.
- You'll be met at the entrance of the building.（建物の玄関でお迎えします。）

G10.2　現在進行形の受動態
現在進行形の受動態もよく用いられます。

例：
- First, the axes of the graphs should be studied in order to determine what sort of information is being represented.（8.2）

G11　分詞
分詞（-ing, -ed）は動詞の意味を堅持しつつ，特定の構文では形容詞としての機能ももちます。現在分詞は現在進行の行為を表します。

例：
- You can produce a spark walking across certain types of carpet in dry weather.（乾燥した空気の中である種のカーペットの上を歩くと火花が散ることがある。）

過去分詞は行為や状態の終了を指します。

例：
- I noticed strawberries eaten by birds.

完了分詞は主節の動詞より以前に起きた行為や事態を示します。

例：
- Having been displaced by a force, the body returns to a state of **static equilibrium** after a few minutes.（力による変位が生じた後，体は数分後に静的平衡状態に戻る。）

G11.1　コンマと分詞
分詞や分詞（句）が文中に現われる際，その前にコンマを打つ場合と打たない場合では意味が異なります。

例：
- The student earning the highest-grade point average will be given a special prize.（最高点平均を獲得した学生には特別賞が授与される。）

特別賞を与えられる学生は "highest-grade point average" を獲得した学生です。ここで分詞句は修飾する名詞の内容を規定（制限）しています。このような場合にはコンマはつけません。

例：
- The apparatus, known as an ohmmeter, is designed to measure resistance.

ここでの器具はオームメーターという名を補おうと補うまいと抵抗計測用に設計されていることに変わりはありません。ここでの分詞句は修飾する名詞の内容を規定（制限）していません。このような場合にはコンマが必要です。このコンマの機能は，関係詞においても同一です。

例：
- I have three sisters, whom I haven't seen for years.（私には姉妹が3名います。ここ数年会っていません。）
- I have three sisters who live in Kyushu.（九州に住む姉妹が3名います。）

2つ目の文は，私には4名以上の姉妹がいることを暗示しています。

G11.2　論文で多用される分詞表現

論文で次のような分詞表現が多用されます。高校時代に by –ing の形を学んだと思いますが，技術英語の論文では by を用いない方が一般的です。本テキストでも多く見られます。

例：
- Students need to learn the basics of presenting their research in a clearly structured format making use of sections and headings so that the information is easy to locate and follow. (1)

G12　動名詞と不定詞

「私は読書が好きです」を英語では "I like reading." あるいは "I like to read." といいます。reading は動名詞，to read は不定詞です。いずれも「～すること」を意味します。両者の共通することと異なるところを確認しましょう。

G12.1　動名詞と現在分詞

動詞であり形容詞の機能ももつ分詞と，動詞であり名詞の機能ももつ動名詞は，共に –ing の形をとりますが，別のものです。

例：
- John is playing the piano.（playing – 現在分詞）
- John is fond of playing the piano.（playing – 動名詞）

G12.2　不定詞の機能

不定詞は名詞句（主語，補語，直接目的語）や形容詞句，副詞句として機能します。

例：
- To see is to believe.
  ＝To see：主語　to believe：補語

- I want to go.
  ＝to go：直接目的語

- I have nothing to say.
  ＝to say：形容詞句。nothing を修飾する

- Some electrons require only little energy to become free.（電子の中には，わずかなエネルギーで自由の状態になるものがある。）
  ＝to become：副詞句。require を修飾する

G12.3　～すること

動名詞と不定詞は共に「～すること」という意味を表します。

例：
- I like to swim. ＝ I like swimming.

G12.4　動名詞と不定詞の違い

動名詞と不定詞には，違いがあります。

(1)　動名詞は繰り返し行われる行為や活動を指すのに対し，不定詞は 1 回の行為を指す

例：
- I enjoy listening to Bach.
  ⇒よくバッハを聴いている

- I went to the nearby city library yesterday to read a novel by Mori Ogai.
  ⇒昨日の一度の行為を記している

(2)　動名詞は前置詞の目的語になれるが，不定詞はなれない

例：
・We're thinking about going for a walk in the woods.
・We look forward to seeing you.

上記例の to は前置詞の to であり，［to ＋ 不定詞］の to ではありません。"with a view to -ing（〜する目的で）" も同様です。

G12.5　動名詞のみを目的語にとる動詞

不定詞は未来に起こる行為を指すのが基本です。

例：
・We decided to go alone.
・We promised to go alone.

動名詞は普段行っていることやすでに行われた行為を指すのが基本です。

例：
・We stopped experimenting on animals.

動名詞と不定詞の意味の違いからは予測できない場合があります。その例に出合ったらその都度覚えていってください。

例：
・Restate the main point for emphasis. This involves paraphrasing or rewriting the topic sentence.

G12.6　how to, when to, whether to, in which to

how to や when to はおなじみの表現ですが，whether to や in which to には戸惑うかもしれません。実例で学んでください。

例：
・A pilot study is a small-scale experiment or series of observations carried out to determine how and whether to launch a larger project（Table19.1）
・It gives readers an overview of the paper and helps them decide whether or not to read the full text.
・When writing a paragraph, you must choose a logical order in which to present information so that one idea leads to another.（1.1.2）
・Another way to show your understanding of the original text is to reorder the information in the sentences while retaining the original meaning. However, deciding which phrases to move or to which position the phrases must be moved is not always easy.（11.2.2）

G12.7　助動詞の働きをする［be to］

いわゆる［be to］（不定詞）という表現があり，いくつかの助動詞の意味を担います。予定や義務・命令，推測，可能などです。

例：
・The gymnast is to make a quadruple somersault.（体操選手は4倍の宙返りをすることにしている。）
・You are to make four straight-line moves.（あなたには4つの直線的な動きをしてもらいます。）
・No new star was to be found.（新星は発見されなかった。）
・Study the factors that have to be taken into consideration when a nuclear power plant is to be built and the need for nuclear energy use.
・It should be clarified whether you wish to do these further studies yourself or whether it is to be left to the wider scientific community to pursue.（6.1）

G12.8 ［to + 副詞 + 不定詞］

to と不定詞の間に副詞を置くことがありますが，論文では避けてください。

論文で使う例：
・I needed to gather my personal belongings quickly.

論文以外で使う例（インフォーマルなニュアンスが強い）：
・I needed to quickly gather my textbooks.
（quickly を強調した表現）

G13 助動詞

助動詞には can, could, may, might, must, shall, should, will, would などとその否定形があります。助動詞の用法は大きく分けて 4 つあります。

(1) 義務や必要性：must, have to, need to, ought to

例：
・Energy must be **transferred** in order to reach equilibrium.（平衡に達するには，エネルギーを伝達する必要がある。）

transfer 伝達する（trans-は「移動」，fer は「運ぶ」の意味をもつ）

・He needs to control the transfer of energy.（彼はエネルギーの伝達を制御する必要がある。）
・You ought to be aware of the situation.（あなたは状況を知っているべきだ。）

> must は不可避の要求や義務を表します。should は可能性や義務，アドバイス，勧めを表します。
> 例：
> ・You must be punctual.（時間を守ってもらわないと困る。）
> ・You should be punctual.（時間は守った方がいい。）

(2) 必要がない：do not have to, do not need to, need not

例：
・An emf device does not have to be an instrument.（emf デバイスは機器である必要はない。）
・The center of mass of an object need not lie within the object.（オブジェクトの重心は，オブジェクト内にある必要はない。）
・He need not have been concerned.（彼は心配する必要はなかったのに。）

(3) 禁止：cannot, must not, may not

例：
・The component of the angular momentum of the system along that axis cannot change.（その軸に沿ったシステムの角運動量の成分は変更できない。）
・The electric field at the surface must not change.（表面の電界は変化してはならない。）
・These copies are licensed and may not be sold or transferred to **a third party**.（これらのコピーはライセンス供与されており，第三者への販売または譲渡を禁じられている。）

a third party 第三者

(4) 可能性：will, should, ought to
ⅰ．100% の可能性

例：
・An instructor's lectures will always be the most valuable learning tools.（インストラクターによる講義は，常に最も価値のある学習ツールになる。）

ⅱ．90% の可能性

例：

・You still should be able to use a conversion table. (それでも換算表を使えるはずだ。)
・He ought to **know better**. (もっと分別があってしかるべきだ。)　　　　　　　**know better** 分別がある
・That surface can enclose no net charge. (その表面は正味の電荷を囲むことができない。)
・The re's likely to be heavy **snowfall**. (大雪になりそうだ。)

iii．50% の可能性

例：
・We could include those equal amounts in the calculation. (それらの同等の金額を計算に含めることができるかもしれない。)
・You may wish to review the discussion. (ディスカッション内容の復習をしておくといいだろう。)

iv．40% の可能性

例：
・The top layers of soil along the slope might slip. (斜面に沿った土の最上層が滑る可能性がある。)

G14　仮定法

　仮定法には2つのパターンがあります。仮定法というと(2)のタイプの文を指すと思いがちですが，(1)の例文の，"he accept our offer" は私たちが希望している（＝実際に起きてはいない）事態であり，仮定法として捉えることができます。"We suggest that he accepts our offer." とはいえません。

(1)

例：
・We suggest that he accept our offer. (我々の申し出を彼には受け入れてほしい。)
・Gauss' law requires that the net flux of the electric field through the surface be zero. (ガウスの法則では，表面を通る電界の正味の流れはゼロである必要がある。)
・It is essential that the resistance of the ammeter be very much smaller than other resistances in the circuit. (電流計の抵抗は，回路内の他の抵抗よりもかなり小さくする必要がある。)

(2)

例：
・If I were you, I would accept the offer. (ぼくが君の立場なら，その申し出を受け入れるんだが。)
・I wish I could fly. (飛べたらなあ。)

G15　単文，重文，複文

　文には3つのタイプがあります。単文（simple sentence）と重文（compound sentence），複文（complex sentence）です。

G15.1　単文

　単文の主要構成要素は主語と動詞です。単文を5つの文型として把握する見方があります。

(1)　S + V

例：
・He died.
　 S　 V

(2)　S + V + C

例：
・She's <u>a worker/busy</u>.
　S V　　　C
・<u>He became a worker/poor</u>.
　S　　V　　　　C

(3)　S + V + O

例：
・<u>She loves science</u>.
　S　V　　O

(4)　S + V + O + O

例：
・<u>He gave me a book</u>.
　S　V　O　O

(5)　S + V + O + C

例：
・<u>She made her students (feel) happy</u>.
　S　V　　O　　　　C

G15.2　重文

　重文は，単文と同じ構造をもつ複数の独立節が以下のように接続詞やコロン，セミコロンで結ばれた文です。コロンとセミコロンの使い分けについては G16.2.1 を参照してください。

(1)　and

例：
・All superfluous things are eliminated **and** accuracy is embraced throughout.

(2)　コロン

例：
・The abstract should include as many key words as possible related to the method and content of the text: these will make it easier to find the abstract in computer searches. (10.2)

(3)　セミコロン

例：
・The stories will contain almost identical information; however, the sentences will be quite distinctive. (11.1)

G15.3　複文

　when や after などの接続詞（従属接続詞）で節と節が結ばれている文を複文と呼びます。

例：
・**When** I was eighteen, I moved to Tokyo.
・I moved to Tokyo **when** I was eighteen.

G16　他の有用な表現

G16.1　ハイフン

　2語以上の単語がハイフンで結ばれ，形容詞の働きをして名詞の前に置かれることがあります。ハイフンを用いた簡潔な表現は，理工系の論文で多く使われています。本書でも随所にハイフンが用いられていますので，それらをノートにまとめるなどして，ハイフンの使用法を習得してください。

例：

・In 1956, Frank Lloyd Wright proposed/recommended/suggested the construction of a mile-high building in Chicago.（9 Exercises）

"a mile-high building" は "a building that is a mile high" を簡潔にした表現です。

・an easy-to-use source（使いやすいソース）
　"The source is easy to use." を名詞句にした表現です。

・a well-honed piece of writing（研ぎ澄まされた文章）

なお，"a specially designed workshop"（特別に設計されたワークショップ）では，"a specially-designed workshop" とハイフンを必要としません。逆に "a mile-high building"（1マイルの高さのビル）を "a mile high building" とはいえません。"a mile high building" では，mile が building を修飾する可能性があり，「1マイルに及ぶビル」となってしまいます。一方，"a specially designed workshop" では specially は副詞ですので workshop を修飾することはありません。specially が修飾するのは designed という形容詞（過去分詞の形容詞的用法）ですので，ハイフンは不要です。

G16.2　句読点
G16.2.1　コロンとセミコロン
コロン（：）は「イコール，すなわち」の意味をもちます。次の例は論文でよく使われますが，(1) から (3) までの終わりには，セミコロン（；）を用います。

例：
・The major points of an entire paper can be summarized in the following order:
(1)　the main goal of the study and the research problem (s) under investigation;
(2)　the fundamental design of the study;
(3)　major findings or patterns identified through analysis;
(4)　a brief explanation of the author's interpretations and conclusions. (8.2)

セミコロン（；）の形式はピリオドとコンマとの合体であり，これらの中間的役割を担います。セミコロンを使いこなすには時間がかかります。『セミコロン』[*1] という，セミコロンを考察した本もあります。本書で使われているセミコロンをノートにまとめるなど，その用法を経験的に習得してください。

*1 ワトソンセシリア著 萩澤大輝・倉林秀男訳 『セミコロン』（左右社, 2023）

例：
・A cause is something that produces an event or condition; an effect is something that happens as a result of an event or condition. (24)

G16.2.2　A, B, and C あるいは A, B and C
2つの書き方がありますがどちらを使っても構いません。なお，A, B, C の A, B, C がそれぞれ長い表現であるときには，B と C の境界をはっきりさせるため A, B, and C とします。

例：
・Difference between suggest, propose, and recommend (9.3)
　　　　　　　　　　A　　　　B　　　　　　C
・Climate change has detrimental effects on livelihoods, habitats and infrastructure. (8 Exercise)
　　　　　　　　　　　　　　　　　　　　　A　　　　B　　　　　C

G16.3　挿入表現
挿入表現を用いることで，文章の流れを単調にならないようにすることができます。

例：
・Trade is not large, however, owing to the costs of transportation. (24.1 (2))
　⇒ "However, trade is not large owing tothe costs of transportation." が元の文

・That, I believe, covers most of the points. (14.2)
　⇒ "I believe that covers most of the points." が元の文。主語と動詞の間に挿入語が置かれている

本書には挿入表現が多くみられますので，それらから使い方を学んでください。

G16.4 倒置表現

高校で，強調表現としての倒置表現を学んだと思います。"I saw John only yesterday."（ジョンに会ったのはほんの昨日のことだ。）の yesterday を強調して "Only yesterday did I see John." と表現します。

例：On no account should the limitations of the research project be hidden. (22.2)

強調の用法としての倒置表現の他に，新しい情報を後ろに置くために用いる倒置表現があります。次の例で説明します。

例：
・It is always a good idea to justify your points of view. Do not simply say "I agree", but rather "I agree because I think that … (explain your reasoning)." Below are some phrases used when agreeing with someone; it is always better to be as enthusiastic as possible when in agreement. (23.2)

これは同意の表現の例を示している文で，同意の表現は新しい情報なので，後置されています。元の文は，"Some phrases used when agreeing with someone are (shown) below;" です。

新しい情報を後ろに置くために用いる倒置表現の習得にも時間がかかります。その例に出会ったらメモして学習し，いつか自分でも使えるようにしてください。

G16.5 動詞と目的語などの特定の結びつき

日本語で「匙を投げる」は文字通りの意味を表すほかに，「あきらめる」という意味があります。これは，「匙を」という特定の目的語と「投げる」という特定の動詞が結びついて，特定の意味を表します。英語には "kick the bucket"（バケツを蹴る）という表現があり，「死ぬ」という意味で用いられています。このような表現を英語では collocation と呼んでいます。co-（共に）locate（置く）という意味です。

例：
・In science, an investigative question must be posed, (16)
・Only one step of the procedure should be addressed in one sentence. (4.1)
・The success of the research project will be measured by how well these aspirations are met. (20 Text)

G16.6 強調の do

強調の do はある文脈の中で用いられます。たとえば，パーティーへの出席を尻込みしている人に対し，"We do want you to come to the party." と，do を使うことで「パーティーに来てほしい」という気持ちを強く表します。do は強く発音されます。

また，以下の Hydropower is ～. の文では，水力発電には汚染物質は含まれないが，それでも環境への何らかの影響を及ぼすという事実を，強調の do（ここでは does）を用いることで表現しています。

強調の do を高校時代に学んでも，自分で使ったことはないかもしれません。次のような例を手がかりに使ってみてください。

例：
・Sometimes definitions need to say what the term is not and is thus called a negative definition. If such a definition is needed, it should always be followed by a statement expressing what the term actually does mean. (20.4)
・Hydropower is non-polluting but does have environmental effects.

G16.7 and と or の否定表現

ド・モルガンの法則（A かつ B の否定は $\overline{A}$ または $\overline{B}$，A または B の否定は $\overline{A}$ かつ $\overline{B}$）の適用例として，次の英語の否定表現があります。

例：
・Don't drink and drive.（飲酒運転はしてはならない。）

⇒ drink and drive（酒を飲んだ後に運転する）の否定は，don't drink（酒を飲まない），or（または）don't drive（運転をしない）

・I don't drink or smoke.（私は酒を飲まないし，タバコも吸わない。）
⇒ drink or smoke（酒を飲むかタバコを吸う）の否定は，don't drink（酒を飲まない），and（また）don't smoke（タバコを吸わない）

・An insulator is a substance that does not transmit heat, sound, or electricity. (20.2)

G16.8　省略表現

たとえば，"I respected Professor Johnson, and my brother <u>respected him</u>, too." の繰り返しの表現である respected him を省き，それを do（does, did 代動詞と呼ばれます）で置き換え，"I respected Professor Johnson, and my brother <u>did</u>, too." あるいは "I respected Professor Johnson, and <u>so did my brother</u>." と表現するのが普通です。

次の例文には異なる種類の省略表現が見られます。省略された表現を補ってみてください。

例：
・Some context to orient the readers who are less familiar with the research topic should be offered and the significance of the research work established. (3.1)

省略された表現を補うと "the significance of the research work should be established" になります。まずは省略表現に慣れ，そして使えるよう努力しましょう。

G16.9　論文でよく用いられる表現

論文でよく用いられる表現をいくつか紹介します。

G16.9.1　account for

account for をみなさんは「説明する」の意味で記憶しているかもしれませんが，以下の例では「占めている」という意味です。

例：
・That is to say, the left half of the system accounts for 10% of the initial intensity. (15.1)

G16.9.2　contribute to

機械翻訳サービス Deep L を使うと，「明瞭さと簡潔さを欠く要因には次のようなものがある。」と訳されます。「貢献する」という基本的な意味からこの用法への意味の広がりを想像してみてください。

例：
・The following factors contribute to a lack of clarity and conciseness: (12)

G16.9.3　be responsible for

以下の文では「A が B をもたらす」という意味です。「責任がある」という訳語ではこの文に対処できません。

例：
・The interference will be responsible for the brightest possible illumination. (24.1)

G16.9.4　respectively

例：
・In an experiment, to determine deformation under force, 15 metals were assessed for rigidity using horizontal and vertical forces respectively. (5.6)

「15 種類の金属を<u>それぞれ</u>水平方向と垂直方向の力を使って剛性を評価した」とは，「15種類の金属の<u>すべてに対し</u>水平方向と垂直方向の力を使って剛性を評価した」ということです。別の例を挙げます。"John, Mary and Nancy went back to their respective rooms."（ジョンとメアリー，ナンシーはそれぞれの部屋に戻った。）は，ジョンはジョンの，メアリーはメアリーの，ナンシーはナンシーの部屋に戻った，という意味です。

G16.9.5　thereby

thereby は「それ（there＝that）によって（by）」という意味です。therefore は「それ（there）ゆえに（for）」という意味です。

例：
・The experimental findings revealed significant changes in the chemical composition, thereby indicating the transformative effects of the reaction.（Table6.2）

G16.9.6　e.g.

「たとえば」を意味し，ラテン語の exempli gratia（＝ for the sake of example）に由来します。

例：
・Using a single verb（e.g. *analyze*）rather than a verb followed by a noun（e.g. *conduct an analysis*）is preferable.（12.2）

G16.9.7　i.e.

「すなわち」を意味し，これもラテン語の id est（＝ that is）に由来します。

例：
・The rest of the paragraph should be devoted to supporting the hypothesis i.e., giving reasons for your opinion.（1.1.1）

G16.9.8　et al.

「その他」を意味し，やはりラテン語の et alii（＝ and others）に由来します。

例：
・That's a fascinating question, but I'm afraid I don't have an answer; however, I believe the answer may be found in the works of Tanaka et al., published last year.（13.6）

G16.9.9　（and/or）vice versa

vice versa（その逆）というラテン語から借用された表現は，and／or vice versa として用いられます。

例：
・If the original sentence is in the active voice, it can be changed to passive and vice versa.（11.2.3）
・four horizontal and one vertical or vice versa（水平 4 本，垂直 1 本，またはその逆）

G16.10　論文と短縮形

論文では don't のような短縮形は用いません。

例：
・Experimental points that do not seem to follow the general trend need to be examined closely.（7.2）

付録W

語源 Word origin

　語源の知識はとても重要です。語の適切な使い方を教えてくれます。また，原義（最初の意味）とその後の意味変化から，その語がこれまでどのように使われてきたかを知ることができるのです。現在使われている，また過去に使われた同義語や関連する語の間の微妙な違いもわかるようになります。また，語源を学ぶと，言語と民族の歴史への理解が深まります。単語の歴史を知ることで，過去において社会が社会に，あるいは民族が民族に及ぼした影響がみえてきます。英単語の歴史の学習を通して，言語と文化が他の多くの言語や民族との交渉を通じて成立したことをみてとることができます。

　英（単）語の歴史を知るうえで重要なことが３つあります。

　(1)　インド・ヨーロッパ語族とグリムの法則
　(2)　ラテン語・フランス語の影響
　(3)　大母音推移

　本章ではW1〜3でこの３つを簡単に紹介し，W4以降は具体例を示します。例文末尾（　）の章・節・項番号は本文と対応しています。

W1　インド・ヨーロッパ語族とグリムの法則

　ロシアを含むヨーロッパの大多数の言語（フィンランド語，ハンガリー語，スペイン北部のバスク語などを除く）は，インドの言語（ヒンズー語）やペルシア語などと，かつて同一の言語（一つの言語）でした。それをインド・ヨーロッパ祖語（Proto-Indo-European）と呼びます。ある時期，ゲルマン民族が大移動を行い，それに伴って，ゲルマン民族の言語（ドイツ語，オランダ語，デンマーク語，アイスランド語，英語など）の特定の子音に変化が生じました。これは他の Proto-Indo-European の子孫の言語とは異なる子音変化です。表 W.1 のように，b, d, g, p, t, k がそれぞれ p, t, k, f, th, h に変化したのです。たとえば，pedal と foot は同じ語源から来ています。この事実を「グリムの法則」と呼びます。グリム童話を収集したグリム兄弟の兄が体系化したので，この名がつきました。

表 W.1　グリムの法則による子音変化

子音変化	例
b → p → f	labial（唇の）→ lip
	pedal（足に関係ある語）→ foot
d → t → th	dental → tooth
g → k → h	cognitive（認知の）→ know
	unicorn（一角獣）→ horn（角）

W2　ラテン語・フランス語の影響

　私たちが英語と呼んでいる言語は，5世紀半ばに今のデンマークの辺りからイングランドに移住した人々（アングロ人・サクソン人・ジュート人）によってもたらされました。これらの人々はアングロ・サクソン人と呼ばれています。

　当時の英語はドイツ語やオランダ語などとよく似ていました。たとえば「私」は，ドイツ語では「イッヒ」，オランダ語では「イック」，英語では「イッチ」でした。

　1066年にイギリスは，フランスのノルマン半島の一諸侯であるノルマンディー侯ウィリアムに敗れ，以降ある時期まで，宮廷ではフランス語が公用語とされたのです。それを受け，大量のフランス語（とその基のラテン語）の語彙が英語に流入しました。ask（本

来の英語）・question（フランス語）・interrogate（ラテン語）のような同義語はその一例です。英語検定準一級程度の語彙力があればフランス語の文を，文法を知らなくてもある程度理解できます。

例を挙げます。「元始に神天地を創造たまへり」という『旧約聖書』の「創世記」第一章第一節はフランス語では "Au commencement, Dieu créa les cieux et la terre." です。この文のうち，英語の commencement は「始まり」という意味で，アメリカ英語では「新しい人生の始まりの日」ということで「卒業式」をも意味します。また，Dieu（フランス語の「神」）と関連のある英単語に deity（神）があります。さらに，créa（創造した）と関連する英単語は create です。cieux（空（複数形））の cie- に l を補うと ciel- となり，そこから i を除くと cel- となります。英語の celestial は「空の，天の」という意味です。「地上の」は英語では terrestrial です。フランス語の terre と関連します。ここで取り上げた英単語はすべてフランス語からの借入語（loan word）です。このように英語は，ラテン語，フランス語の影響を強く受けています。

W3　大母音推移

読者のみなさんが英語を習い始めた頃，英単語のローマ字での発音との違いに悩んだことがあるかもしれません。たとえば，ローマ字で mike は「ミケ」と読みますが，英語で Mike は「マイク」と読みます。i は「イ」だったのに，英語では「イ」でもあり（例：kit），「アイ」でもあります（例：kite）。英語には母音を表す文字（a i u e o）に二通りの読みがあるのです。

make [meik]・time [taim]・house [haus]・feel [fiːl]・fool [fuːl] の発音の変化をみてみると，ある時期までの英語では，make は [maːke]，time は [tiːme]，house は [huːs]，feel は [feːl]，fool は [foːl] と発音されていました。文字と音が一致していたのです。しかしそれらは [meik] [taim] [haus] [fiːl] [fuːl] という，現在の発音に変化しました。これらの母音変化は 15 世紀初頭から二世紀をかけて生じました。この現象は大母音推移（the Great Vowel Shift）と呼ばれています。

最後に，house の綴りについて補足します。house はもともとは hus と綴られていました。u の ou への変化はフランス語の綴りの影響です。フランス語では [uː] の音を ou で綴ります。hus [huːs] → hous [huːs] → hous (e) [haus] と変化したのです。

以下のセクションでは，英単語の語源をいくつか取り上げます。ラテン語・フランス語起源のものが大半ですが，そうでないものもあります（たとえば，otherwise（W5）は借入語ではありません）。なお，英単語の語源を学ぶのに，Online Etymological Dictionary（https://www.etymonline.com/）はたいへんに役に立ちます。

W4　current：走っている

cur- の原義は「走る」であり，-ent（ラテン語の語尾で英語の -ing に相当します）と合わせて「走っている」を原義としてもちます。「電流」は electric current，「現首相」は the current prime minister と表現します。「今走っている電気」が電流であり，「今走っている」を「現在活動している」の意味ととれば，「現職の」という意味が生じます。

例：
・The starting point of our current understanding of it lies in the *law of gravitation* of Isaac Newton. (11.2.2)
・He's currently the president of a major company.

W5　-wise：でないと，さもなければ

ここでの -wise は「聡明な」ではなく，way（方法，やり方）です。otherwise は別のやり方→別のやり方をすれば→さもなければと意味が展開します。

例：
・However, such a huge amount of literature has to be tackled in a correct manner; otherwise（でないと，さもなければ）, it will overwhelm the hopeful scientist at the beginning. (2)

wise ＝ way と理解しておけば，otherwise の他の 2 つの用法や likewise, clockwise なども理解できます。

例：
- I think otherwise. (私は別なふうに（in other ways）考える。私はそうは思わない。)
- He was slightly bruised but otherwise unhurt. (少しあざができたが，他には（＝他の点では）けがはなかった。)
- The wheel is inverted so that it is seen to be rotating clockwise (時計と同じように，時計回りに). (3.2.2)
- He accepted the offer and expected his family to do likewise (同様に).

W6　predict：pre-（前もって）＋dict-（言う）

接頭辞の pre- は「前」，語幹の dict- は「言う」という意味をもちます。

例：
- Readers can use information from a text, relate it to their prior knowledge and use it to make logical predictions before and during reading. (2.1 (2))
- The body of your paper is where you display your research, make pertinent points if you have an argument to make, contradict (反論する) other arguments, and list your references. (14.1 (3))

contradict（contra-：反対のこと　dict-：言う）には２つの意味があります。他の人間と反対のことを言えば「反論する」なのに対し，一人の人間が反対のこと（相反すること）を言えば，それは「矛盾」です。dictator は「言う人」，「独裁者」の意味をもちます。

W7　propel：pro-（前に）＋pel-（押す）

propeller は pro-（前）と pel-, -er から成ります。pel- は「押す」が原義です。プロペラ機は propeller-driven plane といい，「プロペラによって動かされる飛行機」のことです。

例：
- It is proposed that a spaceship may be propelled in the solar system by radiation pressure. (17.3)

compel は「共に押す」という意味をもちます。力を合わせて押せば相当の力となり，強制力や説得力を発揮します。

例：
- Recommendation reports are more overtly compelling（説得力がある）because they provide a final recommendation based on study and data. (29.1)

repellent は「押し返す力をもつもの」で，「防虫剤」です（re- は back の意味をもつ接頭辞です）。

W8　de-：形をこわすこと

対象物を変形することを「デフォルメ」といいます。これはフランス語からきており，英語では deformation といいます。de- は「離れて，分離して」の意味をもつ接頭辞です。「対象物の形から離れる」→「対象物の形をくずす，変形する」と意味が広がります。deforestation は「森から木を分離する，切り倒す」ことであり，「森林伐採，森林破壊」のことです。「脱アミノ酸」は deamino acid といいます。

例：
- To determine deformation under force, 15 metals were assessed for rigidity using horizontal and vertical forces respectively. (5.6)

W9　emit：放出する

e- は ex- と同じで「外へ」，mit あるいは mis は「送る」という意味で，全体で「外に送る，放出する」という意味になります。

例：
- The paths can be seen because of the emitted light.

mission（使命）の原義は「ある目的のために派遣されること」です。ほかに，transmit（送信する）などがあります。

W10　reduce：re-（後ろに）＋duce-（導く）

duce- の原義は「導く」です。conductor は導く人であり，「指揮者」，「バスなどの車掌」の意味をもちます。induction（帰納）は「個々の具体的な事柄から一般的な命題や法則へ（in-）導き出すこと」であり，deduction（演繹）は「大きな一つの大前提から（de-）結論を推論すること」です。

例：
- The functions of discourse markers are quite varied, including contrast, deduction（演繹）, example, addition and summation.（26.4）
- When/As the field is removed, the induced（誘導する）dipole moment disappears.（4.2.2）

W11　sist-：立つ

アシスタント（assistant）の sist- は「立つ」という意味をもちます。as- は ad-（＝to）の異綴語で，assistant は「〜に向かい立っている人，〜のそばに立っている人」が原義です（attraction の at- も ad- の異綴語です）。

consistent（首尾一貫した）は con-（「強意」の意味を表す）と sist（＝stand），-ent（-ing に相当するラテン語語尾）から構成され，「しっかり立っている → ぐらつかない → 首尾一貫した」という意味をもちます。

例：
- It is better to save the readers from exerting any mental effort and at the same time influencing them to look favorably on the interpretation of the experimental data that is consistent（一致している，首尾一貫した）with your own deductions（推論）.（7.4）

insist（主張する）の原義は「〜の中に立っている」であり，persist（持続する）の原義は「いつまでも（per-, permanent）立つ」です。

例：
- It is preferable to revisit a discussion when everyone is fresh rather than trying to persist in getting your point across when everyone is tired.

W12　extract：ex-（外に）＋ tract-（引く）

tractor（トラクター）は「荷物を運ぶ，引っ張る」機能をもっています。tract- は「引く」の意味をもちます。「抜粋，要旨」の extract は，「多くの内容から抜き出された主要なもの」です。ex- は「〜から」の意味をもつ接頭辞です。extract には「エキス」の意味もあります。動詞の extract（/ɪkˈstrækt/）は「引き出す」という意味を表します。他に，contract（契約）は「ある取引を関係する両者が引っ張る（取る）」ことです。

abstract（アブストラクト，抽象的な）の原義は「〜から（本質的なもの）を取り出す」であり，attractive「（自分）に向け（at-）引き付ける」が原義です。distract（気を散らす）は，「（気持ちを）別の方向に（dis-）向ける」ことです。

例：
- Just by reading the first and last paragraph of an article or the first and last sentence of a paragraph, you can usually get a pretty good idea of what the extract is about.（8.3）
- Because it is impossible for an audience to read and listen at the same time, too much writing on the slides will detract from（（重要なもの）から離れた方へ引っ張る，〜を損なう）what the speaker is saying.（13.3）

W13　ject-：投げる

subject には「主語，主題」の他に「臣下」という意味があります。およそ無関係に思えますが，ject- が「投げる」という意味をもつことから考えてみましょう。subject の原義は「下に投げる」です。subject A to B は「A を B の下に置く」です。セクション 29.1 の例で確認してください。「下に置かれた者」が「臣下」です。人間によって論じられる（論じる人の下に置かれる）のが「主題」です。なお，subject は，名詞と動詞の綴りが同

じである時，名詞では前を動詞では後を強く読む，名前動後の一例です。export, import, record などがあります。ただし，report は名詞でも動詞でも「後」を強く読みます。

　reject の re- は back（action に対する reaction の re- です）の意味をもち，reject は「投げられてきたものを投げ返す，拒絶する」という意味になります。

　　例：
　　・The words *propose* and *proposal* in research or a work context signifies something formal or official which requires a formal acceptance or rejection. (9.3)
　　・This is the maximum amount of stress to which the steel wires should be subjected. (29.1)

W14　vent-：来る

　"Prevention is better than cure." という諺があります。「事前に防ぐことは事後の治療に勝る」という意味です。vent- は「来る」の意味をもち，prevent は「前に来る，事前に来る」というのが原義で，「防ぐ」となります。

　convention の原義は「共に（con-）来る（-vent）」です。convention には 2 つの意味があります。人々が集まって行うのが「話し合い，大会」です。また，人々が集まって共に住むと，「しきたり，慣習」が生まれます。

　　例：
　　・There is a conventional（慣習的な）method of writing a conference abstract: the stages of such an abstract are generally subject matter, motivation, research problem, method, results, consequences and conclusions. (10.4)

W15　cede-：行く

　「譲歩する」という意味をもつ concede は cede- が「行く」という意味であることを踏まえれば覚えやすくなります。con- は「共に」ですので，「共に行く」となります。人と人とが共に行く，あるいは共に生きるには，譲り合いが欠かせません。precede（～に先行する）の pre- は「前」を表す接頭辞です。「前代未聞の」の意味をもつ unprecedented は，precedent（先を行くもの。先例。-ent = -ing）→ precedented（先を行かれる。先例がある。）→ unprecedented（先例がない）と，順を追って考えると納得がいきます。

　　例：
　　・When a header precedes（～に先行する）an introduction sentence, you should carefully check it for redundancy. (12.3)
　　・President John F. Kennedy's audacious goal of "going to the moon in this decade" in 1962 inspired the American people to unprecedented（かつてない，前代未聞の）levels of innovation in a variety of fields, not just space exploration. (19 Text)

W16　sum-：取る

　assume という動詞は一見無関係な 2 つの意味をもっています。「仮定する」と「責任を取る，引き受ける」です。sum- は「取る」という意味をもっており，「仮定する」という意味の assume は「いくつかある可能性の中から一つを取る」ということです。

　　例：
　　・Although its share of total energy consumption has declined, oil remains Japan's most important source of primary energy. (24 Text)

W17　he-：つながり

　English Heritage という，イングランドの歴史的建造物を保護する目的で英国政府により設立された組織があります。「遺産」という意味をもつ heritage の he- は「つながり」を意味します。

　論理的な文章を書く際に求められるものに cohesion と coherence があります。cohesion は「文と文の内容がつながっている」ことを指し，「結束性」と訳されています。一方 coherence は「文章全体の内容がつながっている」ことを指し，「一貫性」と訳されます。co- は「共に」という意味の接頭辞です。

例：
・The multiple ways in which pieces of a text are linked together（grammatical, lexical, semantic, metrical, alliterative）are referred to as cohesion.（26）
・Whatever they are working on, they need to create a routine. Times for specific activities should be scheduled, and rigorously adhered to.（20 Text）
・Usually a research process also suffers from some inherent（その論文がもつ，固有の，内在する）weaknesses that may not always be possible to completely eliminate.（22.1.2）

adhere の ad- は to の意味をもつ接頭辞です。adhere to は「～に固着する，～に忠実である」という意味を表します。inherent の in- は「～の内部に」の意味の接頭辞です。

inherent は「～の中につながっている，保たれている」の意味をもちます。他に，heredity（遺伝），inheritance（遺産相続），an adhesive tape（絆創膏）などがあります。

W18 　-philic（～を好む）と -phobic（～を嫌う）

-phile, -phobe は「～を好む人」「～を嫌う人」です。technophobe は「テクノロジー恐怖症の人」です。-philic は形容詞で，その反意語は -phobic です。「交響楽団」は philharmonic を意味し，「ハーモニー（調和）を愛する」が原義です。philosophy（哲学）の原義は「智への愛」です。

例：
・The discovery has been confirmed for both hydrophilic bare silicon surfaces and a hydrophobic silicon substrate.（Model paper）

参考文献 Futher reading

Brieger, N., & Pohl, A. "Technical English: Vocabulary and Grammar" (Heinle Cengage Learning, 2002)

樋口晶彦 (1996) 科学技術英語における受動態の考察―妥当な見解とは何か？―. 工学教育, 44 巻 6 号, 16-20

Wallwork, A. "English for Writing Research Papers" (Springer, 2016)

アカデミックな英語を書く

William Strunk Jr. & White, E.B. "The Elements of Style, Fourth Edition" (Longman, 1999)：英語で論文を書くすべての人が繰り返し学ぶべき古典です。コンパクトな一冊です。

綿貫陽，マーク・ピーターセン『表現のための実践ロイヤル英文法』(旺文社, 2006)：「書く英語」と「話す英語」が区別されています。例文もすばらしいです。

マーク・ピーターセン『日本人の英語』(岩波書店, 1988)『続日本人の英語』(1990, 岩波書店)：著者の理工系大学教員の英文添削体験に基づいた書です。著者は日本文学研究のために来日し，本書をはじめ，日本人の英語学習に役立つ本を，楽しく読めるように工夫しながら，日本語で多数執筆しています。

門田修平，氏木道人，伊藤佳世子『決定版 英語エッセイ・ライティング 増補改訂版』(コスモピア, 2014)：本書で扱った内容を，理工系と文科系の垣根を越えて論じています。たとえば cohesion や coherence，コロンとセミコロンなど，本書で学んだことの復習，確認ができます。

日本語と英語の違い

池上嘉彦『英語の感覚・日本語の感覚』(NHK 出版, 2006)：日本語と英語の違い，たとえば，日本語は be 言語で英語は have 言語といったことについて言及しています。この本を通して英語という言語に親しみをもてるようになるでしょう。著者は日本を代表する言語学者です。

ケリー伊藤『英語ロジカル・ライティング講座』(研究社, 2011)：日本語発想を抜け出し，ロジカルに英語を書くことを目指した書です。100 個の日本語文に対し，それを修正した後で模範英文が示されています。

理工系の英語

小野義正『ポイントで学ぶ科学英語論文の書き方 改訂版』(丸善出版, 2016)：理工系の研究者である著者が，理工系大学での講義をもとに執筆した書です。「英語論文を書くことは才能ではなく技量（スキル）である」と著者は説いています。

志村史夫『理科系の英語』(丸善出版, 1995)：アメリカの大学で研究生活を送った理工系の研究者による新書です。本書により，実践的で役に立つ情報が数多く得られるでしょう。

杉原厚吉『理科系のための英文作法』(中央公論新社, 1994)：副題にあるように，「文章をなめらかにつなぐ四つの作法」がわかりやすく説明されています。著者は数学者です。

英語史・語源

堀田隆一『英語の「なぜ？」に答えるはじめての英語史』(研究社, 2016)：付録G語源で

紹介した「グリムの法則」や「英語の語彙に占めるラテン語・フランス語の多さ」「母音の発音の変化」をはじめ、「動詞の三人称単数形のs」「goの過去形はなぜwentなのか」など、みなさんが英語を習い始めた頃から不思議に思っていたことをわかりやすく説明しています。

寺澤盾『英語の歴史』(中央公論新社, 2008)：「およそ1500年にわたる英語の歴史を辿ってみると、それはまさに波瀾万丈の物語である」の一文に始まる一冊で、「生命科学・工学の進歩」、「英語は国際語としての地位を保ち続けられるか」といったトピックも数多く盛り込まれています。著者は日本を代表する英語学者です。

寺澤盾『英単語の世界』(中央公論新社, 2016)：たとえば、駅の「プラットフォーム」は、ITの世界では「ある装置やソフトウェアを動作させるのに必要な、基盤となる機器やソフトウェア、ネットサービス」を意味するように、文脈によって意味が変化する単語があります。この本では、「多義語と意味変化から見る英単語の世界」が魅力的に紹介されています。

風間喜代三『印欧語の故郷を探る』(岩波書店, 1993)：英語がインド・ヨーロッパ語族に属する、という事実に関心があれば、この一冊に挑んでみてください。

清水建二、すずきひろし『英単語の語源図鑑』(かんき出版, 2018)：103種類の語源を取り上げ、同じ語源から生まれた多くの語について、例文付きでわかりやすく解説しています。たとえば、perfectの-fectは「作る」という意味であり、effect, affect, infect, defectという重要語がその語源を基に説得力をもって解説されています。

清水建二『教養の語源英単語』(講談社, 2022)：「scienceの本来の意味は『知ること』」、「atomは『これ以上切ることができない物質』」といった、興味津々のトピックが盛りだくさんな一冊です。

アカデミックな英単語

東京大学教養学部英語部会編著『東大英単』(東京大学出版会, 2009)：発信型英語力を身に着けるために、eradicate, sequenceなど厳選された280語の使い方を例文を交えてていねいに解説しています。

問題の解答 Answers to questions

Part 1
Chapter 1
Q1 i. c ii. Recent missions to Mars have not turned up definite proof of life.

Chapter 2
Q1 (a) scanning (b) scanning (c) scanning (d) skimming
Q2 (b)

Chapter 3
Q1 i. to make ii. To make
Q2
i. They have introduced computer-guided robots in order to increase efficiency.
ii. The refrigerator was modified so that the energy loss is minimum.
iii. The north pole of the magnet must face the approaching north pole so as to repel one another.

Chapter 4
Q1 i. make, filled ii. flowed iii. driven iv. cooled v. controlled
Q2 i. observe, experiment, predict, investigate, interpret, construct

Chapter 5
Q1 i. analysis ii. developed iii. advancement iv. experimental v. involving
Q2
i. Earth is a little warmer because it is nearer to the Sun than Saturn.
ii. The gold nucleus is much heavier than the alpha particle.

Chapter 6
Q1 i. compiled ii. recorded iii. determine/identify iv. improve v. search for
Q2 i. result from ii. result in iii. result of

Chapter 9
Q2 i. proposed ii. suggested iii. proposed

Chapter 11
Q1 i. (b) shock ii. (c) interest iii. (a) possible

Chapter 12
Q1 citation, communication, conservation development, education, information, introduction, prediction, product, solution

Chapter 14
Q2 i. aim ii. begin iii. problem iv. finally, emphasize

Part 2
Chapter 16
Q1 i. should only be administered ii. should be prepared iii. must always be turned off iv. must never be released v. must never be allowed
Q2 i. science ii. faculty iii. available iv. research v. evidence

Chapter 17
Q1 (c)
Q2 (1) (a), (2) (c), (3) (d), (4) (b)
Q3 i. banned ii. must iii. supposed iv. prohibited

Chapter 18
Q1 i. (b) ii. (b)
Q3 i. seem ii. believed iii. appears iv. might
Q4 i. may have ii. might not have had a mechanical failure iii. cannot have collapsed

Chapter 19
Q1 through, to, first, initially
Q2
i. They turn the orbiter tail first, then they fire the RCS thrusters.
ii. Once the block has been either pulled or pushed away from its original position before being released, it moves in a straight line in simple harmonic motion.
iii. When the displacement is zero, the acceleration is also zero.
iv. As soon as the circuit is finished, the charge starts to flow.
v. After the switch is closed, current flows through the resistance.

Chapter 20
Q1 chemist physicist zoologist engineer geneticist
Q2 i. (b) ii. (a) iii. (d) iv. (c)
Q3
i. A hovercraft is a vehicle which travels across land or water.
ii. GPS or Global Positioning System is an electronic system that uses a network of satellites to indicate the position of a vehicle, ship, person, etc.
iii. An ammeter is an instrument that measures the strength of an electric current in terms of amperes.
iv. Torque is a twisting force which generally causes something to rotate around an axis or another point.

Chapter 21
Q1 i. (d) ii. (a) iii. (c) iv. (b)
Q2
i. Electrons are used to generate a magnetic field.
ii. A laser continuously monitors the thickness of the product.
iii. Curved portions of highway sections are always banked (tilted) to prevent cars from over-turning.
iv. A heat pump is an air conditioner operated in reverse to heat a room.

Chapter 22
Q1 i. hypotheses ii. acquire
Q2 i. misleading ii. Concern iii. Complicated

Chapter 23
Q2 i. (a) I agree ii. (a) I agree iii. (b) I disagree iv. (a) I agree v. (a) I agree. vi. (a) I agree vii. (b) I disagree.

Chapter 24
Q1 fossil fuels, energy sources, plentiful supply, global warming
Q2 i. (b) ii. (a) iii. (c)
Q3 i. result in ii. As a result of iii. Because of iv. Caused by v. due to

Chapter 25
Q2 Certainly, most likely, definitely, probably, cannot possibly

Chapter 26
Q1 i. bps ii. Transmit iii. Webpage

Chapter 28
Q2
i. They will most likely increase funding for robotic exploration of the solar system. (likely)
ii. Low-Earth orbit space travel will possibly become more inexpensive. (possibly)
iii. Scientists will probably not (be able) construct new heavy-lift rockets in the near future.

(probably)

Q3

i. CO_2 emissions will have risen by at least 21% by 2040, according to projections. (rise)

ii. CO_2 concentrations are projected to have increased to 450 ppm by the year 2040. (increase)

Chapter 29

Q2 i. propel ii. Occupying iii. Selecting

Chapter 30

Q2

i. Frictional forces do not act on the objects in a system that is isolated from its environment.

ii. Project Seafarer was an ambitious program to construct an enormous antenna buried underground.

索引 Index

学術論文の書き方を説明する際に使われる語
（vocabulary used in explaining how to write an academic paper）

技術英語で使われる理工系の語
（vocabulary frequently used in technical English）

文法用語
(vocabulary used for explaining grammar)

一般用語
(vocabulary basically used in general English)

［著者紹介］
Sonia Sharmin（ソニア　シャーミン）
2006 年　オックスフォード大学物性物理学科物理学専攻博士課程修了
現　　在　筑波大学数理物質系　助教，Ph.D.（Applied Physics）
専　　門　Magnetic Materials

奥　浩昭（おく　ひろあき）
1987 年　東京大学人文科学研究科英語英文学専攻修士課程修了
現　　在　電気通信大学情報理工学研究科　名誉教授，文学修士
専　　門　英語学
主　　著　「リーダーズ英和辞典（第 2 版)」（分担執筆，研究社，1999)

理工系の技術英語
　　―論文の作成・発表に
　　　必要なスキル―

Technical English for Science
and Engineering:
Writing and Presentation Skills

2024 年 7 月 31 日　初版 1 刷発行

著　者　Sonia Sharmin　ⓒ 2024
　　　　奥　　浩昭

発行者　南條光章

発行所　**共立出版株式会社**

〒112-0006
東京都文京区小日向 4-6-19
電話　03-3947-2511（代表）
振替口座 00110-2-57035
URL www.kyoritsu-pub.co.jp

印　刷　藤原印刷
製　本　加藤製本

一般社団法人
自然科学書協会
会員

検印廃止
NDC 407, 507, 836

ISBN 978-4-320-00617-1

Printed in Japan

■科学一般関連書　　　　　　　　　　　　　　　www.kyoritsu-pub.co.jp **共立出版**

これから論文を書く若者のために 究極の大改訂版 酒井聡樹著

これからレポート・卒論を書く若者のために 第2版 酒井聡樹著

これから学会発表する若者のために 第2版 酒井聡樹著

これから研究を始める 高校生と指導教員のために 第2版 … 酒井聡樹著

100ページの文章術 わかりやすい文章の書き方のすべてがここに … 酒井聡樹著

どう書くか 理科系のための論文作法 ……………… 杉原厚吉著

技術者・学生のための テクニカルライティング 第2版 ……… 三島 浩著

理工系の技術英語 論文の作成・発表に必要なスキル S.Sharmin他著

理数系のための技術英語練習帳 さらなる上達を目指して 金谷健一著

テクニックを学ぶ 化学英語論文の書き方 馬場由成他著

DataStory 人を動かすストーリーテリング ………… 渡辺翔大他訳

あなたのためのクリティカル・シンキング 廣瀬 覚訳

カラー図解 哲学事典 ………………………… 忽那敬三訳

哲学の道具箱 ………………………………… 長滝祥司他訳

倫理学の道具箱 ……………………………… 長滝祥司他訳

証明の読み方・考え方 数学的思考過程への手引き 原著第6版 ……西村康一他訳

「誤差」「大間違い」「ウソ」を見分ける統計学 … 竹内惠行他訳

数楽工作倶楽部 多面体の工作で体験する美しい数学の世界 ………… 廣澤史彦著

この数学,いったいいつ使うことになるの? 森 園子他訳

脳を活かす 空間認知力パズル ………… 黒澤和隆編著

仮説のつくりかた 多様なデータから新たな発想をつかめ 石川 博著

コンピュテーショナル・シンキング …… 磯辺秀司他著

教養としての量子物理 ………………… 占部伸二訳

量子の不可解な偶然 非局所性の本質と量子情報科学への応用 ………… 木村 元他訳

SDGs達成に向けた ネクサスアプローチ 地球環境問題の解決のために 谷口真人編

天気のしくみ 雲のでき方からオーロラの正体まで …… 森田正光他著

竜巻のふしぎ 地上最強の気象現象を探る ………… 森田正光他著

宇宙生命科学入門 生命の大冒険 ……………… 石岡憲昭著

ニュートンなんかこわくない 力学をつくった数学者たち …… 太田浩一著

コンピューター誕生の歴史に隠れた 6人の女性プログラマー 彼女たちは当時なにを思い どんな未来を想像したのか 羽田昭裕訳

コンピューティング史 人間は情報をいかに取り扱ってきたか ………… 杉本 舞監訳

復刊 計算機の歴史 パスカルからノイマンまで ……末包良太他訳

はじめて学ぶ科学史 ………………………… 山中康資著

科学史・科学論 科学技術の本質を考える ………… 柴田和子著

ガリレオの迷宮 自然は数学の言語で書かれているか? ……………… 高橋憲一著

災害対応と近現代史の交錯 デジタルアーカイブと質的データ分析の活用 · 佐藤慶一著

政策情報論 ………………………………… 佐藤慶一著

オムニバス技術者倫理 第2版 オムニバス技術者倫理研究会編

スーパーエンジニアへの道 技術リーダーシップの人間学 ……… 木村 泉訳

コンサルタントの秘密 技術アドバイスの人間学 …… 木村 泉訳

ライト,ついてますか 問題発見の人間学 ………… 木村 泉訳

うるしの科学 ………………………………… 小川俊夫著

図解木工技術 日曜工作から専門まで 第2版 ……… 佐藤庄五郎著

図説竹工入門 竹製品の見方から製作へ ………… 佐藤庄五郎著

デザイン人間工学 魅力ある製品・UX・サービス構築のために ……… 山岡俊樹著

15分スケッチのすすめ 日本的な建築と町並みを描く ……… 山田雅夫著